PENGUIN HANDBOOKS

THE SUMMER STARGAZER

Robert Claiborne is one of the best-known and most-respected science journalists in the United States. A former associate editor of *Scientific American* and later editor of the *Life* Science Library, he has written seven other books: *Time; Drugs; Climate, Man, and History; On Every Side the Sea; The First Americans; God or Beast: Evolution and Human Nature;* and *The Birth of Writing.* He has also edited, with Victor McCusick, M.D., *Medical Genetics* and, with Gerald Weissman, M.D., *The Cell Membrane.* Mr. Claiborne's articles have appeared in such publications as *Harper's, The New York Times, Reader's Digest,* and *Smithsonian.* He brings to this book all his talent for making technical subjects understandable and exciting to the nontechnical reader, as well as his own infectious enthusiasm for astronomy as a hobby.

THE SUMMER

STARGAZER

ASTRONOMY FOR BEGINNERS

ROBERT CLAIBORNE

Star Maps and Drawings by Jonathan Field

PENGUIN BOOKS

Penguin Books Ltd, Harmondsworth,
Middlesex, England
Penguin Books, 625 Madison Avenue,
New York, New York 10022, U.S.A.
Penguin Books Australia Ltd, Ringwood,
Victoria, Australia
Penguin Books Canada Limited, 2801 John Street,
Markham, Ontario, Canada L3R 1B4
Penguin Books (N.Z.) Ltd, 182–190 Wairau Road,
Auckland 10, New Zealand

First published in the United States of America
with the subtitle *Astronomy for Absolute Beginners*
by Coward, McCann & Geoghegan, Inc., 1975
First published in Canada by Longman Canada Limited 1975
Published with revisions in Penguin Books 1981

LIBRARY OF CONGRESS CATALOGING IN PUBLICATION DATA
Claiborne, Robert.
 The summer stargazer.
 Reprint of the 1975 ed. published by Coward, McCann &
Geoghegan, New York.
 Bibliography: p.
 Includes index.
 1. Astronomy—Observers' manuals. I. Title.
[QB64.C57 1981] 523.8'9 80-39576
ISBN 0 14 046.487 5

Printed in the United States of America by
Offset Paperback Mfrs., Inc., Dallas, Pennsylvania
Set in Times Roman

Contents

Preface

Why did not somebody teach me the constellations, and make me at home in the starry heavens, which are always overhead, and which I don't half know to this day?

—Thomas Carlyle

The roots of this book go back nearly ten years—to 1966, when my wife and I rented a cottage near the outer tip of Cape Cod for six weeks. It was enough. The wind in the pines, and the smell of them on a hot, still day, the miles of deserted woodland roads we explored on foot, the surf cresting and pounding on the outer beaches after a northeaster, all did their work, and we kept coming back; our initial six weeks stretched into nine and ten. We continued to rent the same small cottage—where I wrote several dozen articles and parts of five books—until 1973, when growing prosperity, plus a lucky deal on two acres of land, enabled us to build a house of our own a quarter mile away.

Both cottage and house were (and are) surrounded by birds. Bobwhites and towhees awakened us at dawn; mockingbirds serenaded us morning and afternoon—and sometimes, to our irritation, at three A.M.; field sparrows sounded their liquid trills on hot afternoons, and whippoorwills obsessively cried their names through the night. By our second summer I had brought a cheap pair of binoculars and Peterson's *Field Guide to American Birds* and began bird-watching.

Bird-watching led almost automatically to star-watching. The cottage, as it happened, was situated on a hilltop from which, on clear nights, we could see almost the entire dome of

7

the heavens; with binoculars in hand, I began hunting through the night sky for interesting objects even as, during the day, I searched the trees and meadows for interesting birds. In 1971, I bought a small telescope, and discovered how much more there was to be seen.

Once I had learned to operate the instrument with a modest degree of skill—which took considerably longer than I, in my innocent self-assurance, had supposed—I found I had a built-in tourist attraction on my hands. It was a rare visitor on a clear evening who was not eager to peer through the eyepiece at Jupiter and its shifting moons (the planet was prominent in the summer skies of both 1972 and 1973), at the jeweled sapphire-and-topaz of the double star Albireo, at the brilliant star clusters in the constellations Sagittarius and Perseus, or at the mountains and craters of our own moon.

Eventually, the nickel dropped, and my mind disgorged the professional writer's maxim: If you know something about a subject people are interested in, tell your agent. Evidently I knew something about star-watching—though not, as I was to discover, nearly as much as I needed to know—and clearly people were interested in it. Moreover—and this was a key point—there seemed to be no really suitable book that could answer the kinds of questions my friends were asking, or could help them through the difficult beginning problems of stargazing that I had had to solve largely by trial and error. Many of the books I had consulted told me far more than I wanted to know at that stage; some told me a great deal less. None of them provided really clear, simple yet detailed instructions on such crucial topics as finding one's way around the bewildering array of stars in the heavens, or setting up and operating a small telescope. Hence _The Summer Stargazer_.

Why "summer"? The reason, quite simply, is that summer is the only time most people are likely to have the chance to stargaze. In the cities and suburbs where most of us live, the brilliance of the night sky is washed out by the interfering sky-glare from man-made lights. (In the case of a metropolis like Los Angeles, indeed, the glare is sufficient to bedevil astronomical observatories more than fifty miles away.) Yet to see any but the most prominent celestial objects—the moon and the major stars and planets—they must be viewed against a dark sky, and with eyes whose sensitivity to dim light ("dark adaptation") has not been weakened or destroyed by nearby street or house lights.

Another reason, of course, is that in most parts of the United States (or Europe, for that matter) nights in spring or fall—let alone winter—range from chill to frigid, conditions which do not lend themselves to any outdoor occupation that, like stargazing, involves staying relatively immobile for long periods. For anybody, summer is by far the most comfortable time for stargazing, and for most people vacation time away from city or suburb, in the mountains, the seashore or the great open spaces, is the only feasible time.

As it happens, moreover, many of the most interesting celestial objects can be seen most conveniently—*i.e.*, in the hours before midnight—only during the summer months; this applies, for example, to the rich fields of nebulae and star clusters in the constellations Sagittarius and Scorpius. Of the winter's major attractions, on the other hand, most can be viewed during the early morning hours in late August and September. Getting out of bed at four A.M. in August to view the magnificent Orion Nebula is inconvenient for most people, but distinctly preferable, I would say, to sitting outdoors at ten P.M. on a frosty November night.

For all these reasons, then, I have limited this book to the summer skies—primarily those of evening, but in a few cases those of early morning, for readers with the interest and self-discipline to stargaze at those hours.

I have also made something of a point of describing *only what the reader can reasonably expect to see* with the equipment he is likely to possess. Leaf through almost any book written for the amateur astronomer and you will find pages devoted to magnificent photographs of galaxies, nebulae and clusters, taken through the 50-, 100- and 200-inch telescopes of our great observatories. Unfortunately the reader, when he peers through his modest 3- to 6-inch instrument, sees nothing of the sort—usually not even (as he might perhaps expect) a miniature of the photograph. His natural reaction is to feel a bit disappointed, and perhaps even to wonder if he isn't doing something wrong.

The contrast between 100-inch photograph and 6-inch reality is in part due to the nature of telescopes, which (as I shall explain in detail in Chapter 4) are not simply—or even primarily—instruments for magnifying our view of distant objects. But the further truth is that many of the most spectacular photos of celestial objects show things that no man has ever seen, or ever will—not even the few lucky enough to have access to a major observatory. Photographic film, unlike the retina of the eye, can accumulate light over seconds or minutes, thereby building up a visible image of faint objects that even the biggest telescopes cannot make perceptible to the eye. The astronomers at Kitt Peak or Palomar can obviously see innumerable things that the amateur cannot—but the cameras there, in turn, can "see" many things that the astronomers cannot. For all that, however, there is no lack of beautiful and fascinating objects that the amateur *can* see with a small telescope, or even with binoculars, provided he knows where to look, how to look and what to look for; I hope to provide accurate and useful answers to these questions.

Finally, I should stress that this book is exactly what its subtitle says: for beginners. I have not attempted to cover all aspects of amateur astronomy, or even to provide comprehensive lists of all the interesting objects that can be seen with a small telescope or binoculars. There are plenty of books that give this information—though some of them are naturally more useful and readable than others—and I shall cite them at the end of this one, for readers interested in pursuing stargazing beyond the elementary stages. What I aim to provide here is a set of directions whereby the beginner can move with maximum ease through those stages, which are the most bewildering phase of observational astronomy, together with accounts of some—not all—of the interesting and beautiful objects that can be viewed along the way.

Good luck—and good stargazing!

To cite by name the many friends whose curiosity about and excitement at the sky's wonders encouraged me to write this book would be tedious to me and probably embarrassing to them. I owe a special debt, however, to Robert Nelson, now at the Jet Propulsion Laboratory in Pasadena, California, who was kind enough to read the entire manuscript and point out a number of ambiguities and errors. My wife was also invariably helpful—but then, she always is.

Truro and New York, New York, 1980

Chapter 1

HEAVENS ABOVE
Why, When and Where

MANKIND has been stargazing for at least 5,000 years—early written records from Egypt and Mesopotamia testify as much. If we can judge by present-day primitive peoples, our species has been contemplating the heavens for much longer than that: probably since the first men of fully modern type appeared on earth, some 40,000 years ago, and quite possibly still longer. There is no culture on earth today, however simple, that is not aware of the lights that spangle the night sky—gazing at them, naming them, and often making practical use of them. The stars, along with the sun and moon, provided man with his first calendar, as they still do for some tribes of Australian aborigines, perhaps the most primitive cultures on earth. When Arcturus* appears at sunset in the western skies of Northwest Australia, the natives know it is time to hunt for termites; similarly, the appearance of Vega* tells them to begin searching for the toothsome eggs of the mallee hen. The bushmen of southern Africa even credit the stars with a certain control over the seasons. As the dry winter of their year, when food and water are scarce, draws to a close, they chant invocations to Sirius* and Canopus,* brilliant stars of the southern summer (our winter, of course),

*For the meaning and pronunciation of these and other star names, and further information about them, see Appendix B.

urging them to rise higher in the sky, bringing the rains and relative plenty.

As the stars told primitive man when and what to hunt and gather, so they told later men when to plant and harvest; some Indian tribes in Brazil still watch for the day in spring when the Pleiades set just after the sun, which tells them it is time to put in the yam crop. Later, when more complex societies developed, the stars helped men construct the first clocks, their nightly motions across the sky measuring out the night hours as the sun's motions did those of the day. The ancient Egyptian word for "hour," written *wnwt*, was almost the same as that for "starwatcher," *wnwty*.

Still later, mariners learned to use the stars to plot their journeys across the trackless seas. As most people know, the aspect of the heavens changes, not just with the time of night or season of the year, but with the latitude of the observer: the position of a given star with respect to the northern or southern horizon changes as one travels north or south. Canopus may blaze high in the summer skies of the bushmen, but to dwellers in most of the Northern Hemisphere it lies invisible beneath the southern horizon; similarly, Polaris, the North Star, becomes invisible south of the equator. Crafty shipmen in the ancient world learned to keep track of their progress north or south by noting the changing elevation of prominent stars and constellations. In our own era, intrepid Polynesian navigators carried human settlement from island to island for thousands of miles across the Pacific with no more equipment than their outrigger canoes and their wits. And the latter included methods of using the stars as compasses.*

For most modern men and women, the stars have ceased to be of practical use. Printed calendars clue us in to the passing seasons, while clocks tell us the time—though the accuracy of the clocks themselves ultimately depends on the time standards set by various national observatories (such as the U.S.

---

*In the tropics, for reasons too complex to recount here, many stars can be thus employed, since they maintain approximately the same azimuth, or compass direction, for several hours together; in more northerly regions, only Polaris is of comparable use, without fairly sophisticated equipment and calculations.

Naval Observatory in Washington, D.C.), which are periodically checked against star observations. Navigators on ship or plane may still "shoot" the stars—especially our planet's "own" star, the sun—to determine their position, but most of us, if we want to know where we are, will more likely consult a map. Yet for all that, people continue to contemplate the stars, along with the moon and planets. Why?

One reason, surely, is that the night skies appeal to the imagination—as, indeed, they have for centuries, judging from the imaginative and fanciful names that ancient civilizations gave the constellations. Imagination is further stimulated if the contemplative eye is aided by a small telescope, or even a pair of binoculars. To see what Galileo saw with his first crude instrument: the four major moons of Jupiter swinging around their gigantic parent, the rings of Saturn; to view the misty oval that is the Great Nebula in Andromeda and know that it marks a galaxy larger than our own Milky Way whose light began traveling toward us when our ancestors were little more than apes—these are experiences to excite all but the dullest, most deadly "practical" minds.

Not a few celestial objects are of rare beauty in themselves. Binoculars will turn the central region of the constellation Perseus, or the star cluster called the Pleiades, into the semblance of a free-form diamond breastpin; a telescope will do the same for a dozen other clusters of different sizes and shapes. For those who know where to look, the heavens are sown with jewels—sapphire, topaz, aquamarine and rosy garnet; the murky orange-red of Antares and the blazing diamond-white of Vega or Capella.

Perhaps the best way of answering the question "Why look at the stars?" is to take a blanket or deck chair outdoors on a clear, moonless summer night, lie back, and look up for five or ten minutes. If after that you can still ask the question —well, the chances are you'll never know the answer!

Where to Stargaze
Stargazing can be carried out almost anywhere, provided

only that the terrain is reasonably open—thickly wooded areas with no sizable clearings are obviously out—and several miles removed from the nearest town (or village with streetlights). From my own "observatory" I can, on any clear night, see the lights of Provincetown, some seven miles away as the gull flies, stretched along the northwest horizon like a multicolored necklace. But ornamental as they are, I have not a few times cursed them under my breath for blotting out objects I wanted to look at in that direction.

Meadows and pastures are good observation points, especially if they are located on hilltops (but if you are in a "working" pasture, watch out for mushy hazards underfoot!); so are the shores of lakes, or dunes and beaches along the seashore (but dress warmly; bare sand can turn very cold by sundown).

Weather is another important factor in where to watch; the most propitious site will obviously be useless if fog or clouds obscure the sky. Fog is a not uncommon problem in some New England coastal areas, and even more so along the California coast. In the latter region, however, excellent conditions for summer stargazing can usually be found a few miles inland—or a few hundred feet up, where hills or mountains punch through the shallow layer of coastal fog.

The most consistently good stargazing conditions are found in dry or desert regions, including the southwestern, Mountain and plateau states and much of inland California (whose summer climate is almost that of a desert); equally good—if you happen to be in that part of the world—are the Mediterranean lands, whose summer climate is California with a lot less fog, while our own Plains states are almost as good. In these dry regions, the atmosphere is at its clearest, cloud cover is infrequent, and trees—which might interfere with the view—are often sparse. Central and eastern Canada, along with the adjacent American states, are good locations weatherwise, because their air is usually rather dry and hence clear; since their vegetation is more profuse, however, they require more judicious site selection. Elsewhere, the expectable stargazing conditions, which is to say the probable percentage of dry, clear nights, ranges from good (*e.g.*, most of the U.S.

Northeast and much of northwest Europe) to fair or poor (the U.S. Gulf states and Great Britain).

When to Stargaze

Even with a suitable site to stargaze, plus favorable weather conditions, you still can't see everything in the sky all the time. What you can see depends in part on when, during the summer, you happen to be looking (see Chapter 2). More immediately, however, it depends on the positions of the sun and moon, which is to say on the time of day and time of month.

When I began this chapter, I had planned to go into considerable detail on the times of sunrise and sunset in summer. Eventually, I realized that since these vary so much—with the latitude and longitude of the observer, in particular—any attempt to give them in detail would make matters more, not less, confusing. The simplest solution is to consult one of the principal newspapers published in the area you happen to be. Most of these—as well as some smaller dailies and even weeklies—give the local times of sunset and sunrise, moonset and moonrise, as well as the dates of the moon's phases (new moon, first quarter, full moon, last quarter) for the satellite's current cycle or lunar month. A few will even list which planets are visible, and at what times. All this information usually appears on the same page with the weather reports. (Much of it can be found in an almanac, if you happen to have one handy.)

With this information you can easily tell what you can and can't see over the next few nights. It breaks down like this:

1. When the sun is in the sky—*i.e.*, from sunrise to sunset—the only observable objects for the amateur are the sun itself, the moon and the planet Venus, if you know where to look for it. *Observations of the sun itself are extremely dangerous unless you know exactly what you are doing;* do not under any circumstances undertake them until you have read and digested the section beginning on page 154. Daylight observations of the moon are not dangerous, though its features don't

show up as well as they do at night. The only reason for day-light moon-watching applies during its last quarter, at which time it does not rise until the early morning hours; looking at it in daylight saves you from having to stay up late or get up early. On Venus, see page 147.

2. During the half hour or so just after sunset (or just before sunrise) you can observe the moon (if it is visible) and any of the brighter planets (Venus, Mars, Jupiter, Saturn and—sometimes—Mercury) that happen to be in the sky (see Chapter 7).

3. During the next half hour or so, the stars begin emerging—the brightest first, naturally. This is an excellent time for the beginning stargazer to learn the "geography" of the heavens; the fact that only a few stars are visible (as against thousands later in the evening) makes it much easier to pick out the important sky marks and sky patterns described in detail in Chapter 2. If you have just bought a telescope, this period is also a good time to get acquainted with it; there are enough stars to give you targets on which to practice pointing and focusing the instrument, but still enough light to let you see the controls. (Eventually, of course, you will need to be able to operate these by touch, with occasional help from a flash-light.) Twilight, finally, is a good time to watch for artificial satellites—see page 156.

4. Between an hour and a quarter and an hour and a half after sunset, the afterglow fades completely from the sky, leaving only starlight and—depending on the time of month—moonlight. From here on, the moon itself is the decisive factor in what you can and can't expect to see.

a. For about half the lunar month*—roughly, the two weeks beginning with its first quarter and ending with its last quarter—the moon will be in the sky at sunset or shortly thereafter, and will be throwing enough light to blank out all but the brighter stars and planets. This is a good time for brushing up on your celestial geography and the handling of your telescope, as described in 3 above, and also, of course, for observing the moon itself. Two weeks of unrelieved moon-watching may sound rather dull; it isn't, for two rea-

*This is slightly shorter than the calendar month—about 29½ days.

sons. First, unless the weather is exceptionally favorable, you won't be able to observe every night or anything like it; except in dry-to-desert regions, you will be doing well if you get three or four good nights out of seven. Second, and more important, the change in the angle of the sun's rays on the moon brings out different features of it every night, as explained in Chapter 6, so that even on three successive nights you will be seeing a "new face" of the moon each time.

b. The other half of the lunar month—from about last quarter to first quarter—is the ideal time for stargazing in the strict sense. With the moon out of the sky until late evening or even early morning, the sky will be almost black—a perfect background for showing up the dimmest visible objects, including many of the most interesting: star clusters, nebulae, and the few galaxies that can be seen through a telescope of modest size. These dark nights also show up the full glory of the Milky Way, which is a sight with no small merits of its own.

There are two more facts that can help you decide when to stargaze. First, the stars rise four minutes *earlier* every night—which adds up to about an hour every fortnight, so that the stars you can see at midnight on, say, July 1 are the same you will see at eleven P.M. on July 15, ten P.M. on August 1, and so on.* In practical terms, what this means is two things. First, if—like most people—you begin your stargazing as soon as the sky gets dark, start by examining whatever objects of interest are in the *western* sky. Later in the evening—or later in your vacation—they will have dropped below the horizon. Objects overhead or to the east, on the other hand, can be examined at your leisure later in the evening—or in the month.

The second implication of the stars' four-minutes-a-day shift in position is that any descriptions of the heavens, such as I give in the next chapter, are always relative to a particular time of night. For simplicity, I have described the sky as it looks at ten P.M. (Daylight Time) on the first of June, July, August and September. Suppose, however, that in early July you want to examine some constellation or other object that will

*It is probably obvious to you, but I might as well say it anyway: You can never see more than half the heavens at any given moment—the other half is below the horizon.

not appear in the sky at ten P.M. until early August; all you need do is wait until midnight, and refer to the *August* description. If you want to see the early September sky, stay up until two A.M., and so on. The following table of time equivalents should help you keep things straight.

June 1	June 15	July 1	July 15	Aug. 1	Aug. 15	Sept. 1	Sept. 15
10 P.M.	*	*	*	*	*	*	*
11 P.M.	10 P.M.	*	*	*	*	*	*
Midnight	11 P.M.	10 P.M.	*	*	*	*	*
1 A.M.	Midnight	11 P.M.	10 P.M.	*	*	*	*
2 A.M.	1 A.M.	Midnight	11 P.M.	10 P.M.	9 P.M.**	*	*
3 A.M.	2 A.M.	1 A.M.	Midnight	11 P.M.	10 P.M.	9 P.M.	*
4 A.M.**	3 A.M.	2 A.M.	1 A.M.	Midnight	11 P.M.	10 P.M.	9 P.M.
*	4 A.M.**	3 A.M.	2 A.M.	1 A.M.	Midnight	11 P.M.	10 P.M.
*	*	4 A.M.**	3 A.M.	2 A.M.	1 A.M.	Midnight	11 P.M.
*	*	*	4 A.M.**	3 A.M.	2 A.M.	1 A.M.	Midnight

*Sky too light for viewing anything except moon.
**Sky may or may not be too light, depending on local time of sunset or sunrise.

All times are Daylight Saving; if you happen to be on standard time, subtract one hour.

The second useful fact to remember about celestial timekeeping is that the moon rises about fifty minutes *later* every day. (Saltwater sailors, surfers and fishermen will be aware of this through their concern with the tides, which of course are governed chiefly by the moon.) What this means is that a night when the moon sets around nightfall is a good time to see as much as you can in the way of stars and other relatively inconspicuous objects. If you put things off until the following night, you will have to wait fifty minutes after nightfall, on the next night an additional fifty minutes, and by the third night you will be able to do little in the way of stargazing until after midnight. Conversely, if some night, just as you have your telescope set up, the moon rises and dims out the stars, relax and remind yourself that the following night—weather permitting, of course—you will have nearly an hour of looking before moonrise interferes.

Chapter 2

SKY MARKS AND PATTERNS
Finding Your Way Around the Heavens

TO see what you want to see in the sky—and to know what you're seeing when you see it—means in the first place knowing where to look. And this, in turn, means learning to find your way among the initially bewildering array of stars in the heavens. The problem, actually, is not very different from that of learning to find your way around an unfamiliar piece of countryside, or—if you happen to be a boating enthusiast—keeping track of your position in coastal waters. In both cases, you can make good use of a compass and a map or chart, but will also need to learn the position and shape of conspicuous landmarks (or seamarks): the shape of a hill, the direction of a church tower or isolated dead tree, and so on. The skywatcher operates in the same way, using a natural "compass," maps of the sky, and the prominent sky marks he (or she) has learned to identify.

Your compass is, of course, the North Star, Polaris. Almost anyone who has been a boy or girl scout knows how to find this key sky mark; for those who don't, the key is the Big Dipper. The Dipper is almost instantly recognizable. It is the most conspicuous star grouping in the summer sky (and one of the most conspicuous at any season) and, moreover, *looks like its name*—which is more than can be said of most of the constellations.*

*Including the constellation of which the Dipper is technically a part, Ursa Major, the Great Bear, can be seen as a bear only by the utmost stretch of the imagination—and even then the animal must be endowed with a long tail such as no bear on earth ever had.

Before even trying to find the Dipper, however, your first step is to get comfortable, which means sitting or lying; stargazing while standing up is an almost infallible recipe for a stiff neck. A blanket or even dry grass on a hilltop or dune is good; even better is a lightweight, folding deck chair, such as can at this writing be picked up for perhaps twelve dollars (it will, alas, almost certainly be more by the time this book is published). Preferably, your reclining chair should be equipped with arms, which will be invaluable when you get to the point of using binoculars.

Now turn your chair—or yourself—so that you face north. If you have any doubts about where north is, remember the direction that the sun set in, which is northwesterly in early summer, and gradually shifts to westerly in late September. In either case, if you turn your left shoulder in that direction, you will be facing approximately north.

What you are now looking for is a group of seven bright stars (all are of second magnitude*) arranged in the shape of an old-fashioned water dipper. Just after nightfall in early summer the dipper will be standing more or less upright, as if hanging on a wall; later in the season (or later at night) it will have swung around toward the west (your left) until by late summer it will be roughly horizontal, as if lying on a table. But whatever its position, it still looks like a dipper, and will lead you to the North Star. The key is the Pointers, Merak and Dubhe, the two stars that form the front of the Dipper's bowl. Run your eye from Merak (at the bottom of the bowl) to Dubhe (the lip of the bowl) and continue in that direction: The next bright star you come to (also second magnitude) will be Polaris. If you have trouble finding these or other principal sky marks, turn to one of the sky maps on pages 26–27 (June), 36–37 (July), 40–41 (August), 44–45 (September).

Polaris marks the celestial North Pole—the point in the heavens around which all the rest of the stars appear to revolve (actually, of course, it is the earth that is revolving).

*See Glossary.

Another way of saying this is that Polaris lies directly above the earth's North Pole; that is, if you were standing on an ice floe at that chilly spot in the Arctic Ocean, Polaris would be directly overhead.*

For the same reason, Polaris always marks the direction of "true" north—which (as most people know) is the north shown on maps and charts, but not usually the same north shown by a compass.

Since you are not (I assume) standing at the North Pole, Polaris will not be directly overhead, but will be displaced toward the northern horizon. In fact, its elevation above that horizon, measured in degrees, is the same as the latitude of the place you are looking at it from.†

The summer cottage where I am writing this book, for example, lies only about a mile from latitude 42° north, so that Polaris lies 42° above the northern horizon—or would if I could see the horizon; actually, trees block the view somewhat. At the North Pole, latitude 90° north, Polaris (as noted earlier) lies directly overhead—that is, its elevation is 90°. At the Equator, latitude 0°, on the other hand, the star is right on the horizon—meaning it is for all practical purposes invisible. South of the Equator, of course, it is below the horizon and not even theoretically visible.

When you face Polaris, then, you are facing north, with your back to the south, west to your left and east to your right. With Polaris and the Dipper in view, you are ready for your first lesson in finding your way about the sky. Run your eye from the bowl of the Dipper along the curve of its handle and follow that arc until you reach the brilliant star Arcturus (the pun "arc-Arcturus" is, of course, intentional). Golden yellow, it is the brightest star (better than first magnitude) in the northern summer sky, and fourth brightest in the heavens.** Arcturus is your key summer sky mark for learning your

---

*To be absolutely precise, Polaris is not *exactly* at the celestial North Pole, but since it lies only one degree away from it, the difference can be disregarded.

†For readers who have forgotten their geometry a right angle—the angle from directly overhead to the horizon—is exactly 90°.

**It is outclassed only by Sirius, which is the star of the northern winter, and Canopus and Rigil Kentaurus, which are not visible much north of the tropics.

way around the western sky. How you use it, however, will depend to some extent on what time of the summer you are observing (remember that four-minutes-a-day shift in the stars' positions). At this point, therefore, we shall set down four guides to the heavens, keyed to the same time on four different dates—the first of June, July, August and September. The *patterns* of the stars do not change, of course—that is why they are called "fixed" stars—but the *position* of the patterns shifts from east to west as the earth turns (*i.e.*, as the night grows later) and as the earth moves in its orbit around the sun (*i.e.*, as the season grows later).

The Early June Skies

About ten P.M. on June 1, then, Arcturus will be almost directly overhead. Place yourself facing west—that is, with your right shoulder toward Polaris—and look below Arcturus and to your left, where you will see another brilliant star (though not as bright as Arcturus), Spica. Likewise below Arcturus, but to your right, is the moderately bright star, second-magnitude Denebola. Together, the three form an almost perfect equilateral (all-sides-equal) triangle, sometimes called the Spring Triangle.

The three Spring Triangle stars are your guides to three early-summer constellations. Knowing how to recognize constellations—at least the more conspicuous ones—is interesting in itself, but is also important in finding your way around the sky. First, many (though by no means all) constellations form or include distinctive patterns of stars that you can learn to recognize with a little practice, thereby supplying yourself with additional, local sky marks to help you find nearby, but less conspicuous, objects. Second, constellations technically identify specific regions of the sky, so that when we say that some object is "in Leo" or "in Sagittarius" we know, in a general way, where to look for it.

Denebola (the lower right corner of the Spring Triangle) is the second brightest star in the constellation Leo. Run your eyes over the stars to its right and below it (*i.e.*, toward the

How to Use the Sky Maps in This Book

The maps included in the following pages are of two kinds. The four large ones show the entire sky at specific times during the summer; they are intended to guide you in finding your way around the heavens—*i.e.*, locating and identifying particular constellations. For this purpose they have been simplified. Some fourth-magnitude and most fifth-magnitude stars have been omitted, as have virtually all the other objects—clusters, galaxies, etc.—that you will eventually want to look at. The maps show the sky as seen from Lat. 40° N—roughly, the line Philadelphia-Columbus-Denver-Reno-Punta Gorda, Cal. South of this line, southern constellations (*e.g.*, Scorpius) will rise higher above the horizon, northern ones (*e.g.*, Draco) will be lower; north of the line, the reverse will be true. The maps are usable over most of the United States and southern Canada, but readers in the Gulf states and Hawaii may wish to consult the monthly maps in *Sky and Telescope* (see page 174).

The smaller, more detailed maps show "close-ups" of particular constellations—usually a single one, but occasionally two together. These maps include all stars down to fifth magnitude—that is, all you can easily see with the naked eye—with double stars identified as such. They also include all the other objects—including galaxies, nebulae and globular and open clusters—which can be easily viewed through binoculars and/or telescope, along with some not so easy.

To use the large maps, hold them over your head and turn them so that the side labeled "North" faces Polaris; they will then give you an approximate diagram of the heavens at the stated time and date.

So far as the smaller maps are concerned, I have made no effort to orient them in any particular way, though for most of them north is more or less at the top, south at the bottom, etc. Since the orientation of a given constellation changes as it moves across the sky, you may have to rotate the maps back and forth until they correspond with what you are seeing.

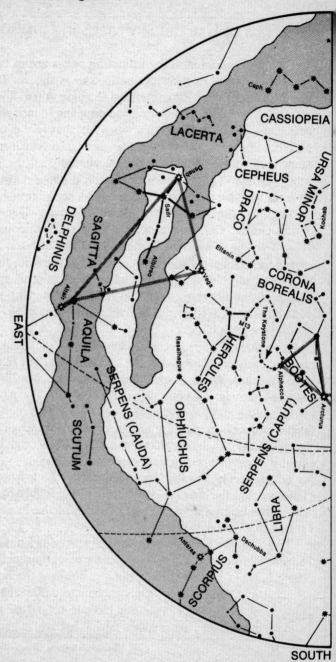

Double Cluster

PERSEUS

CAMELOPARDALIS

Capella

AURIGA

GEMINI

Polaris

Castor

Pollux

URSA MAJOR

Dubhe

Merak

LYNX

CANES VENATICI

Alkaid

LEO MINOR

M44
(The Beehive)

CANCER

Procyon

WEST

Cor Caroli

The Sickle

COMA BERENICES

Denebola

LEO

Regulus

ECLIPTIC

CELESTIAL EQUATOR

SEXTANS

VIRGO

Spica

CRATER

HYDRA

CORVUS

MAGNITUDES

✸ Zero — First

✸ Second

✷ Third

• Fourth

· Fifth

June 1, 10 P.M. (Local Daylight Time)

Instructions for using this map will be found on page 25.

western horizon); the first pattern you will probably see is a trapezoid of four stars, with Denebola on the left and a brighter star (first magnitude), Regulus, on the right. Regulus, in turn, forms part of another grouping arranged like a reversed question mark, with Regulus forming the dot below it; the group is more usually called the Sickle, which it also resembles. All these stars together can, with some imagination, be seen as something like a lion—at any rate, they look more like a lion than Ursa Major looks like a bear. The curve of the Sickle marks the lion's mane, Regulus, its shoulder (or perhaps a forepaw), Denebola, the tip of its tail (a rather short one, for a lion) and the two other stars of the trapezoid, its back. You can now amuse yourself by picking out the stars that mark the animal's muzzle and other paws.

First-magnitude Spica (the lower left corner of the Spring Triangle) is the brightest star in Virgo. Identifying this constellation takes considerable practice, since its stars, apart from Spica itself, are not particularly bright and, moreover, look nothing like a virgin—or any other human female, for that matter. Its more important stars are arranged in a rough "Y," whose bottom, with Spica at the base, points toward Denebola, with the two arms pointing respectively toward Polaris and Regulus.

Arcturus itself dominates the constellation Boötes, named for a mythical Greek hunter and looking even less like a hunter than Virgo looks like a virgin. It can be seen most easily as two curved lines of stars, both extending from Arcturus in the general direction of Polaris; the more westerly of the two almost grazes the tip of the Dipper's handle.

Now turn 90° to your right, so you are facing Polaris. The Dipper will hang high in the sky, above Polaris and to its (and your) left. Almost directly above Polaris you will see two fairly conspicuous stars, which are called the Guardians of the Pole; the brightest is called Kochab. Why the Pole Star should need guardians is something I have never figured out; certainly nobody is likely to run off with it! In any event, the Guardians and Polaris define the Little Dipper, the former marking the front of the bowl and the latter, the tip of the handle, with less conspicuous stars making up the rest of the figure. In contrast with the Big Dipper, the Little Dipper is considerably less "realistic"; its bowl is too wide at the bottom, and the handle curves the wrong way. On the other hand, it looks even less like the Lesser Bear, which is the translation of its official name, Ursa Minor.

Turn now to the east, with Polaris at your left shoulder. High in the sky, and almost due east, is Vega, whose brilliant blue-white light (better than first magnitude) forms a noticeable contrast to the golden yellow of Arcturus. Northeast of Vega (below it and to your left) is first-magnitude Deneb, while southeast of it, and closer to the horizon, is another first-magnitude star, Altair. At ten P.M. in early June, Altair will still be close enough to the horizon to dim its light somewhat, and on a hazy evening you may have trouble making it out. By mid-June (or by eleven P.M.), however, it will have risen high enough to show bright and clear; like Deneb, it shines white with perhaps a slight yellowish cast.

These three bright stars make up the Summer Triangle, which dominates the eastern skies of early summer as the Spring Triangle does the west; even in late summer, when the Spring Triangle has lost two of its three members by nightfall, the Summer Triangle rides high in the sky, as your guide to the constellations in its vicinity. Vega itself dominates the small constellation Lyra, the Lyre, which resembles a miniature equilateral triangle (far smaller than the Spring Triangle) hitched to a parallelogram. Deneb marks the constellation Cygnus, but is most easily seen as the top of the figure called the Northern Cross.* The cross lies on its side, stretching from Deneb, at its head, toward the right (into the Summer Triangle) to the moderately bright (third-magnitude) star Albireo, at its base; the tip of its upper arm lies almost on the line between Deneb and Vega. With a bit more imagination, you can see the Swan (Cygnus in Latin) itself, with Deneb at its tail, Albireo at the tip of its beak, the two crosspiece stars marking the bend of its wings, and two less conspicuous stars

*The Southern Cross, an even more conspicuous sky figure, is not visible much north of the Equator.

at the wingtips, while the star clouds of the Milky Way form the bird's snowy breast.

Altair, like Deneb, marks another celestial bird, Aquila, the Eagle. In early June, however, most of this constellation will be dimmed by haze or below the horizon just after nightfall; we shall have more to say about it in July (see page 35).

By this time, it will be close to eleven P.M. Turn south, with Polaris at your back, and low on the horizon you will see first-magnitude Antares, whose orange-red light forms a noticeable contrast to the golden yellow of Arcturus and the blue-white of Vega. Antares marks the constellation Scorpius—most of which, however, will still be too low to be clearly visible. Note, however, the three moderately bright (second and third magnitude) stars that form a curved line to the right of Antares; they mark the claws of the Scorpion. (Readers in Canada or northern Europe will probably be unable to see Antares at this time; it will be too close to the horizon.)

These, then, are your sky marks in the early summer heavens: Arcturus and its fellow members of the Spring Triangle in the West; the Dipper, Polaris and its Guardians in the north; the Summer Triangle of Vega, Deneb and Altair to the east, and Antares to the south—though this last will be a more prominent beacon later on. A good way of fixing them in your mind is to watch them emerge as the sky darkens after sundown. First to become visible will be Arcturus and Vega, followed by Spica to the southwest and Deneb to the northeast. Altair, being low on the horizon in early summer, will emerge a little later, though in July and August it will appear as soon as (or sooner than) Deneb. Within a few minutes you should be able to see six of the seven Dipper stars; Megrez, where the handle joins the bowl, is noticeably dimmer than the others and will emerge later. Polaris will already have appeared in the darkening sky, followed by its guardian Kochab. Last of the sky marks to become visible will probably be Denebola, which though quite as bright as Polaris lies toward the northwest, where the sun is setting and the sky, consequently, darkens most slowly. With the position and relationships of these stars well in mind, you can fairly easily learn to locate most of the other constellations. Let us see how.

The June Constellations

Beginning with the western sky—other portions of the heavens, as we have noted, can be be left until later in the evening or the month—Denebola guides you to Regulus and the rest of the constellation Leo, as already described on page 28. Beyond Leo (below it and to your right) you will be able to see the bright (first and second magnitude) Twins, Castor and Pollux, the latter being the brighter of the two. They are, as you may have guessed, the two principal stars in the constellation Gemini—most of which, however, will be too near the horizon to see clearly, if at all.

Between Regulus and the Twins lies the constellation Cancer—an inconspicuous group of fourth- and fifth-magnitude stars which, if you can pick it out at all, looks nothing like the crab it is supposed to represent. We shall have a few more words to say about this constellation later on.

Between Arcturus and Denebola lies the constellation Coma Berenices, which is recognizable not because it forms a pattern but because it doesn't; rather, it forms a sort of cloud of quite dim stars—dim enough, indeed, so that you will be able to see few or none of them unless the night is quite clear. This constellation is interesting because it represents the most conspicuous member of a class of objects we shall be discussing and looking for later on: the so-called open cluster of stars. Open clusters are, as the name implies, groups of stars —ranging from a few dozen to a few hundred—lying close together (as astronomical distances go), and usually accompanied by clouds of gas and dust which are faintly illuminated by the stars' light (don't, however, expect to see these clouds with the naked eye or even in binoculars; most of them, in fact, are visible only in fairly large telescopes—or on photographic plates).

An open cluster is a region in the heavens where stars are

being born, presumably from the gas and dust clouds. Astronomers have determined, by methods that need not concern us here, that such clusters consist entirely of relatively "young" stars—meaning that their ages range from a few million to a few tens of millions of years. (But don't expect to see a star born while you're watching; the birth of a star is a considerably more protracted process than the birth of a baby!) Open clusters are also distinctive in that most of them are relatively close to the solar system—the distances range from a few score to a few thousand light-years,* meaning that they are in our "neighborhood" within the great collection of stars known as the Galaxy. For comparison, the more distant objects in our galaxy—notably, those toward its center—lie at distances measured in tens of thousands of light-years, while the nearer galaxies are separated from our own by millions and tens of millions of light-years. (The most distant galaxies, visible only in the largest telescopes, lie at distances estimated—by various complicated methods—in the *billions* of light-years!)

In the northern heavens the three interesting constellations of early summer (apart, of course, from the two Dippers) are Draco, Cepheus and Cassiopeia. Draco, the Dragon, curves sinuously around the Little Dipper. Its tail begins near Dubhe, closest of the Pointers to Polaris, and is defined by a line of third- and fourth-magnitude stars between the Big and Little Dippers. Making a turn of 90° around the bowl of the latter, the figure continues for about the same distance roughly parallel to the Little Dipper, then bends again, this time almost 180°, to terminate in the Lozenge. This is a compact and fairly conspicuous group of four stars lying about midway between the Guardians and Vega; the brightest of them, Eltanin—supposedly representing one of the animal's eyes—is almost as bright as Polaris and Kochab.

Between Draco and the northern horizon—more or less to the right of Polaris—is Cepheus. This somewhat resembles a small hut, with a sharply peaked roof, lying on its side; the peak of the roof points west, toward the Twins.

*For a definition of this term see Glossary and page 71.

Still closer to the northern horizon—it will show up more clearly later in the evening, or in the summer, is Cassiopeia, a conspicuous flattened "W" of three second- and two bright third-magnitude stars. If you have any doubt about where to find it, run your eye along the line from Megrez, at the junction of the Big Dipper's bowl and handle, through Polaris; continuing along this line should just about hit the middle of the W.

The eastern skies have little of interest at this time, apart from Cygnus and Lyra, which we have already met. If you look closely, however, you should be able to make out the small constellation Sagitta, the Arrow—a compact group of four stars lying in the angle of the Summer Triangle marked by Altair.

More interesting is the region of the sky directly overhead —meaning you will have to set your deck chair (assuming you are using one) almost as flat as it will go. If you examine the region near Arcturus, in the general direction of Vega, you should be able to distinguish another sky triangle—almost equilateral, like the Spring Triangle, but considerably smaller, and with a conspicuous star in the middle, which I call the mini-triangle or the Mercedes triangle, because of its resemblance to the emblem of that machine.* Arcturus, being a member of both triangles, forms as it were a link between the two. Of the other three stars in the smaller grouping, two are part of Boötes, while the easternmost is Alphecca, brightest star in the constellation Corona Borealis, the Northern Crown. Its seven stars—mostly third and fourth magnitude except for Alphecca, which is second—are arranged in an inconspicuous but clearly visible three-quarter circle, which gives the constellation its name.

Corona Borealis straddles the line between Arcturus and Vega. Also straddling this line, but closer to Vega, is a figure called—for reasons that will be obvious when you spot it—the Keystone. It does not "square up" with the Vega-Arcturus line, but is tipped toward Vega, so that its top faces the Lozenge at the head of Draco. It is the centerpiece of the con-

---

*I trust the Mercedes people will appropriately reward this free plug!

stellation Hercules, whose other stars—none of them more than third magnitude—extend outward from it in several directions.

Learning all this, unless you are exceptionally quick, will take you several nights (at least!). Once you have a working knowledge of how the June skies are arranged, you can proceed immediately to Chapter 3, which will tell you how to use your eyes and binoculars to spot less conspicuous objects. If, however, you are stargazing later in the summer, read on.

The July Skies

For early July stargazing, your first step is to reread (or read) the preceding section on June. The July skies are not, of course, identical with those of June, but large portions of them are. Though most people know this, the point will probably bear repetition: The position of the stars does not change, or at any rate not in several lifetimes—which is why they are called "fixed" stars; what does change—from hour to hour and from month to month—is the particular portion of the heavens you can see. Thus once you have learned the pattern of a particular constellation or star grouping, the only additional thing you will need to know is how to recognize that pattern as it shifts to different portions of the sky. But the Spring Triangle, the Summer Triangle and all the rest of your sky marks don't change shape, nor do their positions shift with respect to one another. And since I don't want to redefine all these groupings and constellations for every one of the four summer months—it would make for a cumbersome and tedious book—you will need to have their appearance and location firmly fixed in your mind.*

At ten P.M. on July 1, then, † the Spring Triangle will still

---

*If you find yourself getting confused, or failing to recognize a given name in the text, a good tip is to turn to the index and look up the references to it; the one in **boldface type** (as a rule, the first) will tell you precisely what it is and where to find it.

†Or, as explained earlier, at eleven P.M. on June 15.

dominate the western sky. The Twins, however, will have vanished and so, almost certainly, will Cancer, whose dim stars will be unable to make shift against the horizon's haze. Arcturus will have shifted toward the west, while its place near the Zenith will have been taken by Corona Borealis (later in the evening, or the month, by the Keystone).

Facing north, Polaris will of course be where it was (its very slight shift in position can be detected only with a telescope). The Big Dipper, however, will have swung around until its bowl lies almost due west—*i.e.*, to the left—of Polaris, while Cepheus lies almost directly to the right. Below Polaris, the bright W of Cassiopeia will now be clearly visible.

Looking east, the three stars of the Summer Triangle will now be well up in the sky, as will their attendant constellations, Lyra, Cygnus and Aquila. The last is a "new" constellation—*i.e.*, invisible or barely visible at ten P.M. in early June. Its principal stars—third and fourth magnitude, except for Altair—can be seen most simply as a rough triangle made up of two other triangles.

To the left of Altair, and below Sagitta—which lies about one-third of the way from Altair to Deneb—you should now be able to make out a tight group of five fourth-magnitude stars which form the constellation Delphinus, and which in fact bear a certain resemblance to a leaping dolphin: four of the stars, arranged in a diamond, form its body with the fifth suggesting its tail.

To the south, Antares and the "claws" of Scorpius should be clearly visible to all but the most northerly stargazers, and its neighbor in the zodiac, Sagittarius, will be pushing up over the horizon to the left. Sagittarius is most conveniently observed during the early evening in late July or early August, but since it happens to be a constellation containing many interesting objects I mention it here. If your vacation ends in mid-July, or if the moon will be shining later in the month, Sagittarius is well worth waiting up until midnight for. The constellation contains no very conspicuous stars (comparable to, say, Antares or Regulus) and is therefore not all that easy to spot. When I first saw it, its principal (second and third) mag-

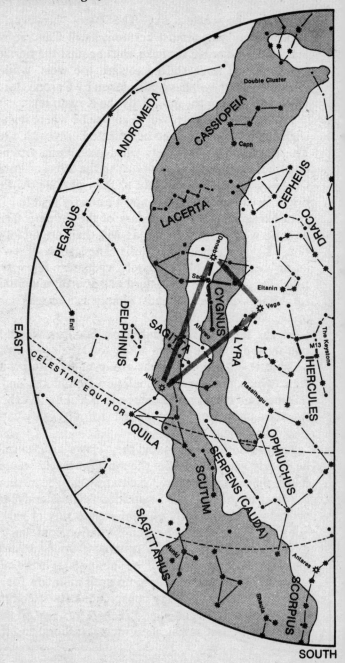

NORTH

Capella

AURIGA

CAMELOPARDALIS

LYNX

URSA MINOR

Kochab

URSA MAJOR

Dubhe
Merak

LEO MINOR

Alkaid

Cor Caroli

The Sickle

CORONA BOREALIS

CANES VENATICI

Algieba

Regulus

LEO

BOOTES

COMA BERENICES

ECLIPTIC

WEST

Arcturus

Denebola

SERPENS (CAPUT)

VIRGO

CRATER

Spica

CORVUS

LIBRA

Dschubba

HYDRA

LUPUS

MAGNITUDES

✸ Zero — First

✹ Second

✴ Third

• Fourth

· Fifth

July 1, 10 P.M. (Local Daylight Time)
Instructions for using this map will be found on page 25.

nitude stars suggested—perhaps because of my long interest in natural history—two angelfish, those crescent-shaped denizens of tropical waters.* You can also see the stars as two bows and arrows—which is what the right-hand group is in fact supposed to represent (seeing the Archer who is doing the shooting requires more imagination than I possess). Or, by taking some stars from both these groups, you can form a figure sometimes called the Milk Dipper; it is shaped much like a smaller version of the Big Dipper, but with only two stars in the handle (see drawing on page 66).

Early July (or late June, for that matter) is also notable because the Milky Way is now high enough above the horizon's haze for its full beauty to be visible. Beginning at the northern horizon below Cassiopeia, it arches across the eastern sky, passing through that constellation, grazing one corner of Cepheus, and just before reaching Deneb splits into two streams.† One follows the general line of the Northern Cross, skirts Lyra, grazes Aquila and peters out at the edge of Ophiuchus, another of the not-very-conspicuous constellations (it lies about halfway between Sagittarius and Corona Borealis). The main stream of the Milky Way passes through Sagitta and Aquila to end on the southeast horizon, in Sagittarius, though it has at this point broadened sufficiently to take in part of Scorpius as well.

As most people know, the Milky Way is in fact an enormous collection of stars so distant that they appear to our eyes as luminous clouds; only a telescope will reveal any of them as individual bodies. Its dimly glowing clouds are the more distant portions of the Galaxy (which is Greek for "milky way"), a flattened disk containing perhaps one hundred billion stars of which our Sun is one—a rather unimportant one. Nor is our galaxy the only such collection of stars in the heavens; powerful telescopes have revealed millions,

---

*Any fish you may see in Sagittarius have nothing to do, of course, with the fish of the constellation Pisces, which have little resemblance to fish of any species—and are, in any case, invisible at this season.

†The "split" is actually an illusion, caused by dense clouds of gas and dust which block off this portion of the Milky Way from our view. Powerful telescopes reveal a few stars behind these clouds, and there can be no doubt that there are far more which are totally blacked out.

most of them so distant that they appear only as dim patches of light. From photographs of the nearer galaxies, we know that most of them are in fact not precisely disk-shaped; some are almost spherical, while others resemble a pinwheel, with their stars arranged in arms spiraling out from an almost spherical center. No such spiral arms can be seen in our own Galaxy, since we are viewing it from inside, but by carefully measuring the distances of various parts of the Milky Way, astronomers have determined that it does indeed have just such a structure, and have even mapped parts of it. Unfortunately much of our Galaxy cannot be mapped or even seen, since our view of it is blocked by the clouds of gas and dust mentioned earlier. For all that, even what we can see of the Galaxy is impressive.

When you gaze up at the Milky Way, it is worth reflecting that you are in fact looking at literally billions of stars—most of them quite as important, in their own neighborhoods, as Vega or Arcturus—or our own sun.

The August Skies

Moving forward another month, we find that at ten P.M. on August 1 the Spring Triangle has almost ceased to be a triangle: Both Spica and Denebola will be so low on the horizon that whether you can see them at all will depend on the local time of sunset and will in any case require a really clear night. This is especially true of Denebola, which is both dimmer than Spica and closer to the sun's setting point. Arcturus, however, will still dominate the western sky, though it will now be almost halfway from the zenith to the horizon, with Corona Borealis and Hercules with his Keystone following in its wake. Vega will lie almost exactly at the zenith, with Deneb and Altair, of course, somewhat to the east.

Looking south, Antares and the "claws" should be clearly visible near the horizon—unless the night is hazy—and watchers south of Maine and Canada should be able to see, a little below them and to the left, Shaula, a conspicuous second-magnitude star marking the "tail" of the Scorpion. Still far-

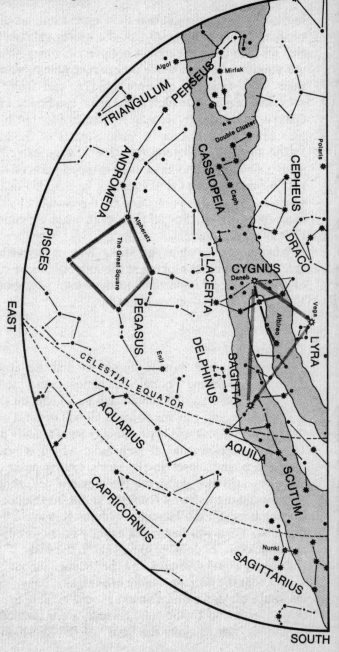

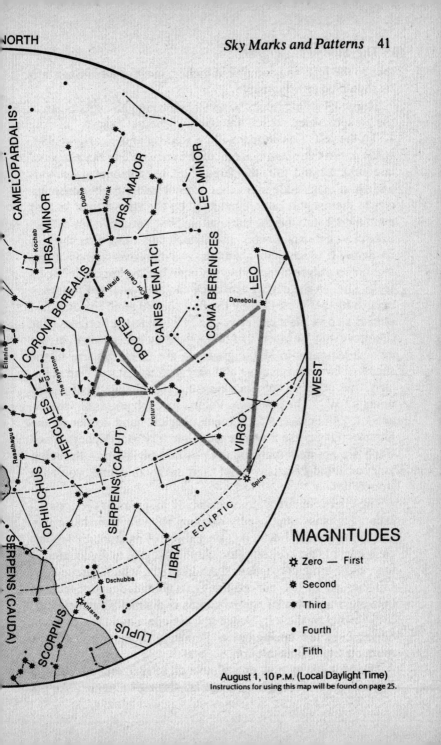

MAGNITUDES

☼ Zero — First

✱ Second

✸ Third

• Fourth

· Fifth

August 1, 10 P.M. (Local Daylight Time)
Instructions for using this map will be found on page 25.

ther to the left, and somewhat higher, most or all of Sagittarius should be clearly visible.

The most interesting changes, as you might expect, are in the eastern skies, which are now two hours "ahead" of July 1. To the left, Cassiopeia will be well up and clearly visible, while almost due east lies one of the outstanding sky marks of late summer and fall, the Great Square, a group of second- and third-magnitude stars that are not particularly conspicuous in themselves but are made so by the almost total lack of even moderately bright stars within the "square" they define. Like most celestial figures, the Square falls somewhat short of its name: It is not quite a square, and moreover stands at this season on one corner, giving it somewhat the appearance of a squat kite. In the early evening of early August, seeing it may be still further confused by the fact that Algenib, its dimmest star, is also its most easterly—*i.e.*, closest to the horizon—and therefore may be blurred or blotted out by haze. However, we have a virtually infallible guide to the Great Square in the shape of two conspicuous and unmistakable stars. One is Polaris, the other, Caph, marking the right-hand tip of Cassiopeia's "W." The line from Polaris to Caph points almost exactly to Alpheratz, which forms the left-hand corner of the Square. (Of course it works the other way too: Alpheratz and Caph are pointers marking the position of Polaris—though if you can't find Polaris without their aid by this time, you're in trouble!)

The Great Square is conventionally assigned to the constellation Pegasus—supposedly marking the wings of that mythical flying horse—but in fact only three of its stars are technically part of that constellation. Alpheratz is actually located in the neighboring constellation Andromeda, which can be seen fairly easily as two curved lines extending from Alpheratz to the left; the lower of the two lines is distinctly the brighter. Still farther to the left, between Cassiopeia and the northeast horizon, is the constellation Perseus—though this will be more clearly visible later on.

The Milky Way now arcs almost directly overhead, passing from Perseus through Cassiopeia, through Cygnus, Aquila and Sagittarius and disappearing below the southern horizon in the region around Shaula in Scorpius.

The Late Summer Skies

The late August and September skies see the disappearance of one of our major sky marks. By ten P.M. on September first, the Spring Triangle has given up the ghost, with two of its three stars—Spica and Denebola—below the horizon.* Arcturus, however, still blazes its golden beacon across the far western sky, with Corona Borealis and Hercules above it. The Summer Triangle now lies toward the west, except for Deneb, which is almost at the Zenith. To the south, Scorpius too will have vanished, and Sagittarius will be partly lost beneath the southwestern horizon. To the east, Pegasus, Andromeda and Cassiopeia will ride high in the sky, with Perseus somewhat lower to the northeast; the Milky Way will arc diagonally across the sky, from southwest to northeast. Near its northeastern end you may already be able to see Capella, a brilliant, yellow-white star almost exactly as bright as Vega. Capella dominates the constellation Auriga, which in a couple of hours (or in late September) will be well clear of the horizon; for its description, see page 76, which will also clue you into some other early-fall treats in this region of the heavens.

---

*You should, however, be able to see them earlier in the evening; by this time, sunset will come early enough so that you can start stargazing at nine P.M. or even a bit earlier.

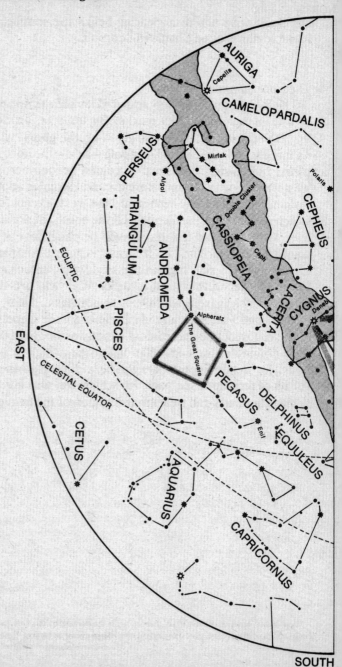

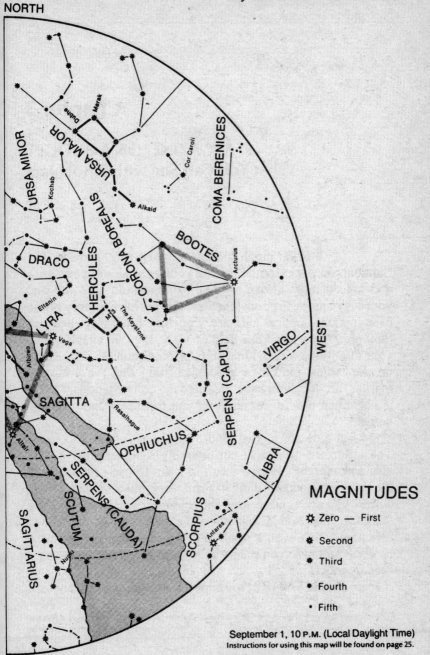

NORTH

URSA MINOR
URSA MAJOR
Dubhe
Merak
Kochab
Alkaid
Cor Caroli
COMA BERENICES
DRACO
HERCULES
CORONA BOREALIS
BOOTES
Eltanin
M13
The Keystone
Arcturus
LYRA
Vega
WEST
VIRGO
Albireo
SERPENS (CAPUT)
SAGITTA
Rasalhague
OPHIUCHUS
LIBRA
Altair
SERPENS (CAUDA)
SCUTUM
SCORPIUS
SAGITTARIUS
Nunki
Antares

MAGNITUDES

✿ Zero — First

✹ Second

✳ Third

• Fourth

· Fifth

September 1, 10 P.M. (Local Daylight Time)
Instructions for using this map will be found on page 25.

Chapter 3

STARGAZING ON A BUDGET
What You Can See with Binoculars

To get much fun out of stargazing, some sort of "optical aid" is essential. Using your eyes alone, you can—and indeed must—learn to identify important stars and find your way among the constellations. In this introductory phase of stargazing, optical aids are a positive hindrance; because they greatly narrow your field of view, they inevitably focus your attention on individual stars or other objects rather than on the larger, patterned groups of stars that are your sky marks. But once you have "learned the constellations," there is not much else of interest that you can see without optical help.

Optical aids are of two kinds: binoculars and telescopes.* The latter are of course considerably more expensive, and also considerably more difficult to learn to use—though both the expense and the difficulty are very much worth coping with if you are at all serious about stargazing. As a beginning stargazer, however, you may well hesitate to make the kind of investment that even a small telescope entails; if so, binoculars will still give you plenty to look at and think about for a season or two.

Even should you wish to shoot the works immediately on a

---

*Binoculars are sometimes called "field glasses," but this is incorrect. Field glasses are an earlier (and less powerful) version of binoculars, and are seldom encountered these days outside an antique shop.

telescope, binoculars are still essential; in fact, trying to use a telescope without first having worked over the sky with binoculars is rather like taking up gourmet cookery before you know how to fry eggs: it can be done, but not easily. Nearly all the objects you will want to examine with a telescope (at least in the first couple of seasons) are also visible through binoculars, though of course more dimly and in less detail. Examining them through binoculars helps to fix their position in your mind (and eye), especially with reference to the smaller, less conspicuous sky marks you will need to know if you are to find them at all, with or without a telescope. It is for this reason that I have gone into a good deal of detail on how to locate many objects which, through binoculars, do not possess much visual interest: Unless you have looked at them that way, your chances of getting to see them through a telescope are small.

About Binoculars*

A pair of binoculars is essentially two tubes, with a lens (or lenses) at each end. (They also contain an arrangement of prisms for reflecting the light through a crooked path, but this doesn't change the basic principle; its purpose is merely to make the instrument more compact.) At the larger end are the "objective" lenses that collect and concentrate light, exactly as a reading glass or "burning glass" can concentrate sunlight so as to set fire to a piece of paper or a tuft of dry grass.†

At the smaller end of the tubes are the eyepieces, which magnify the concentrated light image collected by the large end. By concentrating light—*i.e.*, intensifying it—binoculars enable you to see objects too dim to be perceived, or perceived clearly, by the naked eye. By magnifying the image, they enable you to "resolve"—that is, to see as separate images—things that would otherwise be fused or blurred togeth-

---

*If you already own a pair of binoculars, you may want to skip this section.

†Now do you see why you should never, except under very special conditions (see page 154), look at the sun through binoculars? Imagine the rays from a burning glass focused on the interior of your eyes. . . .

er into one—notably, for example, many features of the moon, and also the components of some double stars.

Any pair of binoculars will have stamped somewhere on it a set of numbers such as 7×35 or 10×50. The first of these tells you how much the instrument will magnify what you are looking at—or, putting it another way, how much "closer" it will bring a given object. Thus 7×35 binoculars, focused on an object seventy feet away, will make it appear only ten feet (70/7) feet away, while a 10×50 instrument will "move it up" to seven feet (70/10). The second number tells you the diameter, in millimeters, of the objective lenses and thus measures their light-gathering power; obviously the more light they gather, the fainter the objects you will be able to see. Most binoculars also include another figure—say, 378—which gives the *width* of your field of vision (usually in feet) at some standard distance (usually 1,000 yards); the field will of course be wider at greater distances, narrower for closer viewing. In star-watching, however, this figure is unimportant and can be disregarded.*

Most people tend to assume that the important thing about binoculars is their magnifying power.† For most ordinary, earthbound uses there is something to this view, since the objects one looks at are generally of fair size, and clearly illuminated. In star-watching, however, you will be looking at objects that are usually dim and almost invariably small, so that the important thing about your binoculars will be their light-gathering power—which, as already noted, depends on the diameters of the objectives—and (to some extent) their resolving power, which, for rather complicated optical reasons, also depends on the objective diameter. Indeed it is almost a rule of stargazing that the more you magnify with a given instrument, the less you are likely to see.

By all odds the commonest variety of binoculars—and also among the cheapest—are 7×35s. Not so many years ago, one

---

*Indeed even for other purposes it is of only minor importance, with the possible exception of bird-watching and sports—and if you're sitting so far from the sporting action as to need binoculars, you'd probably enjoy the game more on television!

†They also make this assumption about telescopes—see page 81.

could pick up an adequate Japanese pair of such binoculars for under $20; nowadays you should count on spending at least $25. Dollar for dollar, however, they are still the best value you are likely to find. Buying a 6 × 30 instrument might shave the price a little, but the loss in both light-gathering and resolving power would more than wipe out the saving.

Somewhat superior for stargazing are 7 × 50s. The larger objectives give you more than twice the light-gathering power of 7 × 35s, since it depends on the area rather than the diameter of the lenses.* Binoculars of this size are sometimes called "night glasses" and are used as such by sailors, since they will often "bring in" dim objects—say, a distant buoy—that are literally invisible to the naked eye. Their disadvantage is that they are more costly than 7 × 35s (about double), hard to find, and somewhat heavier—not very important if you expect to use them for stargazing alone, but a consideration if you also plan to employ them for purposes such as bird-watching, which involve carrying them about.

Still more expensive than 7 × 50s are 10 × 50s, which for stargazing—or most other purposes—are a poor buy, for several reasons. First, neither their light-gathering nor their resolving power is any better than those of 7 × 50s, since both, as we have seen, depend on the objective diameter, not the magnification. In looking at the moon, for example, 10 × 50s will show the details of its surface larger, but they won't show any more details. A second disadvantage is that the higher the magnification, the greater the influence of what I call "shake." Even with both arms braced (by the method I shall describe in a moment) it is impossible to hold binoculars perfectly still—and the higher their magnification, the more the image you see will jiggle back and forth across your field of vision.† A final disadvantage is that 10 × 50s are even larger and heavier than 7 × 50s.

(If you happen to have a camera tripod, *and* the special fixture needed to mount binoculars on it, you can largely cure

---

*Specifically, light-gathering power varies as the *square* of the diameter, so that the proportion between a 35-millimeter and a 50-millimeter objective will be as 1,225 (35²) to 2,500 (50²).

†At 10× magnification, a "shake" of only 1/50 inch will move the image nearly 1/4 inch!

the shake problem—but only at the cost of raising another: Unless you are a master contortionist, you will be unable to use tripod-mounted binoculars to view objects at the zenith, or anywhere near it.)

Most expensive are what are called "zoom" binoculars; these include a special lever with which you can increase the magnification within certain limits. These are described by figures such as 7–14 × 35, meaning that they have 35-millimeter objectives and can magnify the image anywhere from 7 to 14 times. Zoom binoculars also come in such sizes as 7½–15 × 40 (I have such a pair), 8–16 × 50, and I daresay other combinations. While they have certain advantages compared with fixed-magnification instruments, these are not as important as you might think—the more so in that shake makes the maximum magnifications unusable. The maximum useful magnification with a hand-held instrument is about 10× or less so that advertised maximums of 14× or 15× are just that—advertising. Buy a cheaper pair, and put away the money you save toward a telescope.

Whatever type you buy, make sure they are "center focus" (most are), meaning that the focus is controlled by a milled wheel set between the two tubes, so that both can be focused simultaneously; focusing each tube individually is a nuisance.

Adjusting and Using Binoculars*

Not everybody's eyes are the same distance apart, and binoculars are designed to take account of this: The mounting is hinged in the middle, so that by bending it the barrels can be brought closer together or farther apart. Adjust them to suit your own eye distance—*i.e.*, so that you get a full image with each eye—and then note the small scale engraved on the mounting below the focusing wheel, which will probably range from 60 to 70; whatever number registers on this scale is your personal adjustment for eye distance. Noting this num-

---

*Again, you may want to skip this if you're accustomed to binoculars.

ber will enable you to quickly regain this adjustment if somebody else has been using the binoculars, or if (as is likely) you
must bend them slightly to fit them back into their case.

Not everyone has both eyes of equal strength, and binoculars take account of this too. Note on one of the eyepieces
(usually the right) a series of figures, probably something like
−2, 0, +2; the end of the eyepiece can be rotated so that a dot
on it lies opposite one of these figures. If you habitually use
glasses for distance vision, your corrected vision is almost
certainly the same for both eyes, so you can set this scale at 0.
If, however, you have reason to believe that one eye is noticeably more nearsighted or farsighted than the other—without
glasses, if you don't use them for distance, with them if you
do—you can adjust the binoculars as follows:

1. Select some object at a fair distance—say 50 feet. Look
at it *only* through the *non*adjustable (probably left) eyepiece;
you do this by covering the objective of the other tube with
your hand or a piece of cardboard. Adjust the center focusing
wheel until the image is as sharp as you can get it.

2. Now look at the same object only with the other eye,
again using your hand or cardboard. This time, however,
leave the focusing wheel alone, and rotate the movable eyepiece until you get the best focus. Note the position of the
dot, which again marks your personal adjustment (it too will
probably have to be changed if somebody else uses the instrument).

(If, as I do, you habitually use eyeglasses for distance vision, do *not* remove them when using binoculars. Your field
of vision may be slightly restricted—if the eyepieces don't include some arrangement for use with glasses, as they often
don't—but the convenience of not having continually to remove your glasses and replace them far outweighs this disadvantage.)

Spotting a visible star or other visible object with binoculars is simplicity itself. Simply look directly at it with the
naked eye, and then, without moving your head, bring the binoculars up in front of your eyes; they should be pointing di-

rectly at the object. With only a little practice, you ought to be able to hit your object first crack at least four times in five.

The most comfortable—and efficient—way of using binoculars to stargaze is in a lightweight deck chair, *preferably with arms*. Arrange the chair so that it faces the part of the heavens you want to examine, and angle the back so that you can look without undue neck craning. Then sit down, and adjust yourself in such a way that when your binoculars are in seeing position your elbows are resting on the arms of the chair; this will give you a comfortable "mounting" for your instrument, and—even more to the point—a firm one that will hold shake to a minimum.

A handy added piece of equipment is a flashlight that throws a red beam. These can be bought quite cheaply, but an ordinary flash will do well enough if you can cover the business end with a piece of red cellophane held by a rubber band. Red light (as was discovered during World War II) does not destroy the "dark adaptation" of the eyes, so that with a red flashlight you can consult this book, or a star map, out of doors without temporarily impairing your ability to see what it describes.

Even more essential than the flashlight in preserving dark adaptation is proper placement of your chair (or other observation site). Unless you are looking at the moon or one of the planets, which are bright enough to make dark-adapted vision unimportant, you should try to find a spot in which any other light will be minimized or eliminated. Best is a site well away from any house with lighted windows, since even the splash of light on the ground outside can impair your night vision. If you can't find such a site, sit on an unlit side of the house or, if this is impossible, put your back to the light. In general, you can expect that after leaving a lighted room it will take your eyes at least fifteen minutes to fully sensitize themselves to starlight, and any time you look at even a moderately bright light or lighted area this process is set back. The eyes' increasing power to perceive dim objects as they achieve full dark adaptation is one of the facts behind the stargazer's maxim: "The longer you look, the more you see."

Seeing What You Can See

The "circumpolar" constellations—those immediately surrounding Polaris (the Big and Little Dippers, Draco, Cepheus) can be seen at any time during the summer (or, for that matter, during the year). Nonetheless, it is worth starting with them simply for practice.

Begin with the Big Dipper. Note with the naked eye the star Mizar, second from the end of the handle. If your eyes are reasonably good, you should be able to see its fainter companion, Alcor, which lies on the side away from the bowl, and slightly above the line of the handle. If you can't see Alcor, look a little away from Mizar, and both stars should become clearly visible.* For rather complicated physiological reasons, the center of the eye's retina—the part that receives the image when you are looking directly at an object—is less sensitive to dim light than other portions of the retina. Hence shifting your eyes just enough to make the image you're trying to make out fall a little off center is a good trick for seeing any faint object on the threshold of vision.

Now try your binoculars on the two stars: Alcor will come in bright and clear—brighter than it was, and also considerably more separated from its companion, which will of course also look brighter. This is an important point to keep in mind in learning to use either binoculars or telescope: the *relative* brightness of stars does not change, but their apparent brightness does; second-magnitude Mizar seen through binoculars looks very much like first-magnitude Deneb seen with the eye alone.

The Dipper also provides a handy scale for judging distances in the sky. These are measured in degrees (°) and the fractions of a degree called minutes ('); for very close objects (such as the components of double stars), the fractions-of-fractions called seconds ("). To remind any readers who may have forgotten their high school math, there are 360° in a circle and 90° to a right angle—*e.g.*, the distance from the zenith to the "true" unobstructed horizon; 60' make one degree, and 60", a minute.

*If you *still* can't see it, maybe you need glasses!

Note that we are talking here about celestial *distances,* not celestial *positions*—two concepts that are a fertile source of confusion for the beginner. Distances—the apparent separation between two objects—are measured, as just noted, in degrees, minutes and seconds.*

Celestial positions, on the other hand, are given by two figures, corresponding to the latitude and longitude which define positions on earth. Celestial "latitude," called declination, is measured in the same degrees, minutes and seconds used to measure distance. Celestial "longitude," called right ascension, is measured in *hours and minutes*—which are *not* the same minutes used for position and declination.†

At this point, however, we are talking *only* about position; the reason we are taking the trouble to express it in degrees is that—as you will quickly discover—it helps immensely, when describing how to find some inconspicuous object, to be able to say that it is so many degrees from some clearly visible sky mark. As it happens, the Pointers of the Dipper, Merak and Dubhe, are almost exactly 5° apart, while the two stars forming the top of the bowl, Dubhe and Megrez, are about 10° apart. Using these two "rulers," you can begin to estimate distances from one sky mark to another, first with the naked eye, then with binoculars.

In the latter case, you first need to determine the width of your binocular field of vision, which you can quickly do by examining the Pointers through them. If you can just jockey *both* Pointers into the field, then its width is obviously just about 5° (this is the case with most binoculars); if the two of them fit with room to spare, your field will be slightly over 5°– 5½° or 6°. Try to estimate it as exactly as you can.

With this information, the value of your binoculars in finding inconspicuous objects will be greatly enhanced. Suppose, for example, you are told that object X lies 3° from sky mark A; this means that if you get A into your binocular field,

---

*Distances in the sense of how far a given object is from earth are, of course, something else again; *they* are measured in (depending on the amount of distance) miles, astronomical units, light-years or parsecs; for further information, consult these terms in the Glossary.

†For further clarification of these concepts, see page 107, and the Glossary entries under DECLINATION and RIGHT ASCENSION.

you should—if necessary, by moving the instrument so that *A* passes around the edge of the field—be able to get *X* into the field along with *A*. In fact, if you can't do so, you are probably not looking at *A* but at some other star.

Larger distances can be measured by the process called "bridging." Suppose, let us say, you are told that object *Y* lies 7° north of sky mark *B*. Focus on *B*, then shift your binoculars slightly north (toward Polaris) so that *B* lies at the very edge of the field. Note some other star lying at the exact opposite edge; this measures off 5°.* Now turn your binoculars still farther toward the north so that this second star has moved right across the field to the position originally occupied by *B*. Object *Y* should now be 2° away, meaning that it will be near the center of the field. You can, of course, repeat the bridging process to measure even larger distances, but the accuracy falls off somewhat with each step—for which reason I have tried to avoid directions involving long "bridges" of this sort.

Another way of "tuning up" with your binoculars is to examine some of the brighter stars, noting in particular their color contrasts, which the optical aid will often make easier to perceive. Arcturus, Antares and Vega provide an especially clear contrast, from golden yellow to reddish orange to bluewhite. Spica and Regulus, on the west, are, like Vega, bluewhite, while Altair and Deneb, in the Summer Triangle, are almost pure white. Contrast golden-yellow Dubhe with its fellow Pointer, blue-white Merak, and both of them with yellow-white Polaris. Note, however, that all these contrasts are rather subtle—by no means in the class with the glaring color differences in (say) a neon sign. Bear in mind also that haze and the earth's atmosphere itself can distort the color of stars close to the horizon, giving them a yellowish or even reddish cast just as they do with the rising and setting sun.

Now is also the time to begin accustoming your eye to judging brightness through the binoculars. As already noted, this is measured in "magnitudes," the rule being the *lower* the magnitude, the brighter the star; the brightest objects, in fact,

*Assuming, of course, that your field is in fact 5° across.

actually have negative magnitudes.* The planet Venus, which apart from the sun and moon is the brightest object you will see in the sky, varies in magnitude from −4.4 to −3.3 (for the reasons governing this variation, see page 147). The planets Jupiter and Mars also have negative magnitudes, though the latter, at its farthest from earth, drops to magnitude +1.6, considerably dimmer than the brightest stars. Sirius, brightest of the stars, has a magnitude of −1.4, while the dimmest stars visible to the naked eye on a really clear, dark night will have magnitudes of 5 or 6; binoculars will reveal stars down to magnitude 8 or 9. Magnitudes can, of course, be expressed with more decimal places than I have bothered to give. For our purposes, first magnitude means anything between 0.50 and 1.49, second magnitude, from 1.50 to 2.49, and so on.

The summer skies contain numbers of stars that can be used as a scale for estimating brightness. Arcturus and Vega are both almost exactly 0.0 magnitude—meaning that any summer object you see that is brighter is either a planet or an artificial satellite (see page 156). Spica and Antares are close to 1.0 magnitude, as is the slightly dimmer Pollux, brightest of the Twins. Castor, the other Twin, lies almost on the borderline between first and second magnitude, though its magnitude of 1.58 technically makes it the latter. Alkaid, at the tip of the Dipper's handle, and Dubhe, brightest of the Pointers, are both close to magnitude 2.0, with Polaris only slightly dimmer. Merak, the other Pointer, Phecda, which with Merak forms the bottom of the Dipper's bowl, and Eltanin, in the Lozenge of Draco, all lie on the borderline between 2.0 and 3.0—meaning, of course, that all brighter stars are second magnitude or better, dimmer ones, third magnitude or more.

For the dimmer stars, a good guide is the four stars, arranged roughly in a rectangle, forming the bowl of the Little Dipper. Kochab, brighter of the Guardians, is about second magnitude (actually, 2.24) while its fellow Guardian is about third; continuing clockwise around the bowl, the other two stars are fifth and fourth magnitude respectively. Also fourth

---

*For a technical definition of "magnitude," see Glossary.

magnitude are the two stars of the Little Dipper's handle lying between the bowl and Polaris.*

It is useful to get into the habit of comparing stars you are examining with "reference" stars of known magnitude. Until you have had considerable experience, judging a star by its own immediate neighbors can be misleading; if it is surrounded by fainter stars, it may look brighter than it is, while if it is close to a brilliant star like Vega it may look dimmer than it is. There is no reason, however, to get compulsive about judging brightness; its purpose is merely to help you identify particular stars with more certainty.†

If, for example, you are trying to locate a star described as second magnitude, but find yourself looking at one which is only as bright as a known third- or fourth-magnitude star, you are obviously looking in the wrong place.

As you practice judging distances and brightnesses, you can begin looking around the sky for other objects of interest. Of the circumpolar constellations, only Cepheus will yield much to binoculars. The constellation, as noted earlier, resembles a squarish hut with a high-peaked roof, lying on its side. Just after nightfall in early summer, it lies almost between Polaris and the northern horizon, but rotates around the pole as the season progresses until by September it lies between Draco and the eastern horizon, but much closer to the former.

Having located Cepheus, fix your attention on the third- and fourth-magnitude stars that mark the corners of its "base" (the latter may be hard to pick out, since two other stars of about the same magnitude lie close to it); they are about 7° apart. To the right of the line formed by these two stars, and forming a flattish triangle with them, lies the fourth magnitude star Mu-Cep**; it lies about 3° from the upper— and brighter—of the two "base" stars. This object is often

---

*Note that in judging brightness, as in judging colors, stars close to the horizon don't look "normal," appearing dimmer than they actually are.

†And also, perhaps, to impress your friends; it is rather fun to remark casually that such and such a star is second or third magnitude.

**For the meaning of this nomenclature, see Appendix B.

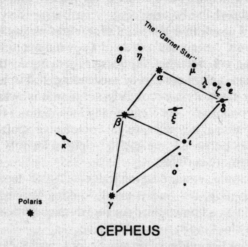

CEPHEUS

The symbols used on this map are identified in Appendix A.

called the "Garnet Star," which is another prime example of a misleading astronomical name. I searched for this star over several summers with absolutely no success—because I expected it to be garnet-colored (*i.e.*, dark purplish-red). While writing this book, I finally pinned down its position—and found that it is in fact pink, almost salmon-colored. But if the name is confusing, the star itself is pretty, even through binoculars, and more so in a telescope.

Another interesting object in Cepheus is Delta-Cep, the lower and dimmer of the two base stars, which is a short-period variable. Every 6.4 days, its light peaks at magnitude 3.8, and about 2 days 17 hours later drops to its minimum of 4.6. The difference in brightness is not easily seen, but can be spotted with luck by comparing the star with the little group of fourth- and fifth-magnitude stars nearby; if it appears as bright as or brighter than the brightest of these, you have caught it near its peak. You will not be able to observe the following minimum, since it will occur during the daytime, but if you can see the star at its *next* minimum, just about four days later, it should look then noticeably dimmer than its neighbors.

Having warmed up, so to speak, on the circumpolar constellations, which can be seen at any time during the year, you can now turn your attention to other star groupings. As usual, we begin with those visible only during the early summer, then work our way east through those visible during the entire season, finishing with the constellations visible—at least during the early evening hours—only in late summer or early fall.

The most westerly early-summer constellation is Cancer, which early in June can be seen for an hour or so after nightfall. Facing west, it will appear as an inverted "Y" of five fourth-magnitude stars; the base of the "Y" straddles the line between Regulus and the Twins, while one of the arms points almost straight down toward the horizon. (You may have trouble picking out the star at the tip of this arm, since even on June first it lies very close to the horizon.)

Now look at the star marking the fork of the "Y." About 2° to its right, you may, if your eyes are good and the night is

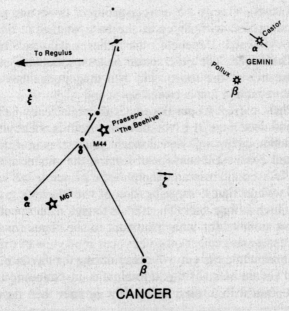

CANCER

The symbols used on this map are identified in Appendix A.

very clear, be able to spot a small, hazy patch of light. More likely, you will need binoculars, which will reveal it as a cluster of several dozen stars. (The total number in the cluster is actually more than 300, but most of these are visible only in a sizable telescope.) This is the Beehive, more formally known as Praesipe, the Manger. It does not in fact much resemble either object, though it might be likened to a swarm of bees —but the same could be said of most clusters. The Beehive, like Coma Berenices, is an open cluster, but its stars are arranged much more compactly, and are also considerably dimmer—the reason being, as you may have guessed, that they are farther (about twice as far) away.

Moving eastward, we come to Leo and Virgo, neither of which, however, is of much interest—except, perhaps, to those born under their signs. Leos should have no difficulty recognizing "their" constellation by the sky marks given on page 24; Virgos will have more difficulty. Its brightest star, Spica, is of course unmistakable, since it forms part of the Spring Triangle; the rest of the constellation can be divided into two parts. One is a Y-shaped group of third- and fourth-magnitude stars, with Spica at the base of the Y; one arm stretches toward Denebola, the other northward toward Coma Berenices. The other more or less noticeable stars in Virgo are five of the fourth and fifth magnitude; they lie on and to the east of a line between Spica and Arcturus.

The open cluster Coma Berenices should already be familiar to you—see page 31 if you have any doubts about where it is. Your binoculars will reveal many more stars in it, though the naked eye is still best for observing the cluster's overall shape. This contrast is an example of a principle we will return to several times: magnification of the heavens can conceal as much as it reveals. The trick is to pick a degree of magnification suitable for what you want to see. Thus for Coma Berenices, *no* magnification gives you a good view of the cluster's shape, while the (say) 7× magnification of your binoculars will reveal most of the individual stars. Stepping up the magnification with a telescope to 50× or 100× will show you only a few more stars—and will totally obscure the cluster as a cluster.

Boötes, the "home" constellation of Arcturus, is, like Leo and Virgo, of little interest. For those interested in picking it out, its principal stars—except for Arcturus, of third and fourth magnitude—are arranged in a roughly kite-shaped figure; Arcturus marks the kite's narrow base, while the fourth-magnitude star at the top of the kite lies about midway between Alphecca, in Corona Borealis, and Alkaid, at the tip of the Dipper's handle.

Corona Borealis shows little of interest to binoculars. We might note, however, that despite the close grouping of its stars it is *not* a cluster. Its components, though close neighbors to the eye, in fact lie at quite different distances from earth and, moreover, are moving—very slowly—in quite different directions, so that in a mere 50,000 the constellation's resemblance to a crown will be totally lost.

Moving east again, to Hercules, the Keystone can guide you to one of the most interesting easily found objects in the sky. If your eyes are good and the night is very clear and dark, look along the right side of the Keystone; about two thirds of the way to the top, you may see—perhaps by looking away slightly—a small, hazy patch resembling a blurred star. Binoculars will make it clearly visible, as a round, glowing patch looking like nothing so much as a tiny ball of absorbent cotton. This is the famous cluster M-13.* To the eye, it is hardly an impressive object—far less so, certainly, than Coma Berenices or the Beehive. You may find it more impressive, however, when you realize that you are in fact looking at something like one million stars!

M-13 is what is called a globular cluster—something much different from an open cluster. Globular clusters are, in the first place, much bigger than open clusters—something like 10 times the diameter, and 1,000 times the volume. Their component stars are much more numerous; where those of an open cluster are numbered by dozens or hundreds, even the small-

*The *M* stands for Messier—see Appendix A.

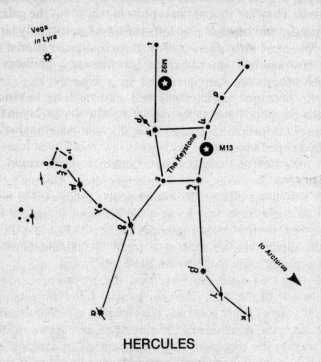

HERCULES

The symbols used on this map are identified in Appendix A.

est globular clusters contain thousands of stars—the largest of them, hundreds of thousands. They are also, however, generally much farther off than open clusters—even M-13, a relatively nearby one, is over 20,000 light-years away—while open clusters with few exceptions lie within a few hundred or few thousand light years from our solar system. Whereas open clusters are found in the flattened, spiral arms of our galaxy, globular clusters are closer to the galactic center, being arranged around it in a loose sphere sometimes called the galactic halo. Astronomers have also turned up a few so-called "tramp" globular clusters that have no apparent connection with our galaxy (or any other), making them the only known exceptions to the rule that stars are found in galaxies, not the immense spaces in between. Finally, globular-cluster stars are not young but old—so far as anyone can tell, as old as the galaxy itself. They are thought to represent regions of the galaxy in which, for unknown reasons, stars formed in specially large numbers—large enough so that their mutual gravitational attraction was sufficient to keep them together for the billions of years that the galaxy has endured.

If your imagination is working well, you might try to visualize what the heavens would look like if our own sun lay in the midst of such a cluster: *every single star you can see,* including those you must now strain your eyes for, would be at least as bright as Arcturus or Vega—and many of them would be much brighter!*

Continuing east from the Keystone, we come to the constellation Lyra, dominated by diamond-bright Vega. To the naked eye, and more clearly through binoculars, Vega forms part of a tight equilateral triangle of stars; binoculars show

*Isaac Asimov's classic science-fiction story "Nightfall" is set on a planet lying in just such a cluster.

The symbols used on this map are identified in Appendix A.

one of these as double. Some people claim to be able to see both members of the pair without binoculars—to which my only comment is that their eyes must be really remarkable. Through a telescope, this double star is revealed as not just a double but a "double-double" (see page 103).

The third star in the triangle is also a double, easily seen through a telescope but only with difficulty (if at all) through binoculars. Like the "double-double," it is a true *binary* star, meaning that its components lie very close to, and revolve around, one another. Several other stars in Lyra are "visual doubles" in binoculars, meaning that they are double only to the eye, their component stars actually lying at quite different distances. Perhaps the clearest way of explaining the difference is that a true binary would appear as such no matter what part of the galaxy it was viewed from, while a visual binary appears double merely because of the angle from which we earthlings happen to be viewing it.

Lyra is also worth looking at with binoculars because of the rich fields of stars that surround it; this is because it lies close to the plane of the Milky Way. When we look at most parts of the sky, we are looking out of the Galaxy, through a cross section of its thin disk of stars. Looking in the general direction of the Milky Way, however, we are looking *into* the disk, not out of it—meaning that we see far more stars. The Milky Way itself, as we know, contains even more stars, though few if any of them will show up as such even in binoculars.

Moving on to Cygnus, so clearly marked by the Northern Cross, the star fields become even richer. Two open clusters in Cygnus are worth looking for, but difficult to spot because of the very richness of the star fields in which they lie; the problem is rather like trying to identify a particular grove of trees in the midst of a forest. For what the information is worth, however, the first cluster, M-29, lies a little over a degree below and to the right of Sadr, the second-magnitude star at the center of the Cross. Look for it in binoculars as a somewhat "grainy" spot of light surrounded by a region noticeably darker than the star field around it. The second cluster, M-39, though somewhat brighter, is even harder to distinguish from its background. It lies about 7° to the left of, and slightly below, Deneb, forming almost a straight line with Sadr and Albireo, at the foot of the Cross. Neither of these clusters, incidentally, looks very different from a globular cluster through binoculars; a telescope, however, clearly reveals that they are typical open clusters of a few score or few hundred stars.

Aquila, lying southeast of Cygnus, is only partly (and probably dimly) visible during the early evening in June. By July, however, it is well up in the sky by nightfall and continues clearly visible for the rest of the summer. Note the two third- and fourth-magnitude stars lying on either side of brilliant Altair and forming almost a straight line with it. These Companions of Altair, Alshain and Tarazed, together with that star, should just about fit within the binocular field, meaning that they, like the Pointers, are about 5° apart and thus provide another "ruler" for judging star distances in this region of the sky. Aquila contains a fairly conspicuous open cluster, lying

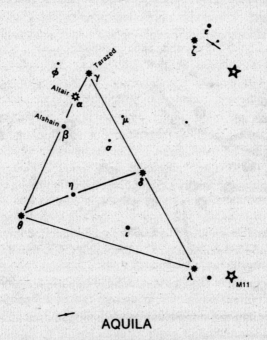

AQUILA

The symbols used on this map are identified in Appendix A.

near one of the points of the large triangle into which most of the constellation's principal stars can be fitted. About 1° 30' (1½°) to the right of the third-magnitude star marking the triangle's lower right corner lies a fourth-magnitude star, and about two degrees to *its* right is the cluster, M-11; all three objects fit neatly into the binocular field.

Scorpius is another July constellation, since unless you happen to be observing from fairly far south—southern California, say, or Georgia—it will be only partly visible in the early evening hours during most of June. However, if your vacation falls in June, wait up for it—it's worth the trouble. The constellation, along with its neighbor in the zodiac, Sagittarius, lies not only in the plane of the Milky Way but in the direction of the Galaxy's center—meaning that the two star groups include a specially dense concentration of interesting objects.

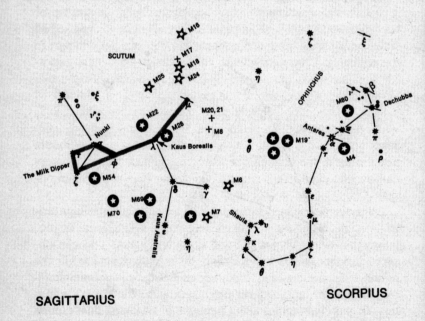

SAGITTARIUS

SCORPIUS

The symbols used on this map are identified in Appendix A.

You can begin with your sky mark Antares, interesting to the eye because of its orange-red color and to the mind because of its nature. It is what is known as a red giant star, relatively cool in temperature—only about 3,500 degrees at its surface, as against the 50,000 degrees of the hottest stars—but enormous in size: The entire earth's orbit would fit inside it, with lots of room to spare.

A couple of degrees to the right of Antares lies a third-magnitude star; about midway between and slightly below them lies the globular cluster M-4, visible through binoculars if the night is clear. Another globular cluster, M-80, lies between and slightly above Antares and Dschubba, the central of the three "claw" stars, but unless the night is exceptionally clear, and Antares well up above the horizon, you will probably miss it even with binoculars, since it lies at about the limit of visibility (eighth magnitude).

Later in the evening (or the month) is a good time to observe Shaula, the bright second-magnitude star at the tip of the Scorpion's tail. The hook-shaped tail itself, composed of second- and third-magnitude stars, can be seen only in part in the central and northern states, but Shaula itself is easily spotted because of its companion, a conspicuous third-magnitude star less than a degree to the right. Nearly 5° to the left of Shaula, and a little above it, lies the rich open cluster M-7, while the somewhat more distant (*i.e.*, dimmer and more compact) cluster M-6 lies about the same distance *above* Shaula and a little to the left. With a little manipulation, both clusters can be jockeyed into the same binocular field with Shaula itself.

An even richer region lies in and directly above Sagittarius—visible in the early evening beginning in late July; it, too, is worth waiting up for if June is the only time you can observe it. Here the Milky Way is at its most brilliant, while the nearer stars are crowded together like daisies in a June meadow; we are, in fact, looking directly toward the Galaxy's center—though we cannot see it because of blocking dust clouds in space. Judging from photographs of other galaxies, if it were visible it would be one of the most spectacular sights in the heavens—an enormous, glowing cloud of packed stars and clusters. Indeed, the estimated density of stars in the Galaxy's central region makes it almost equivalent to a single enormous open cluster—several thousand times the size of those we can see.

Though the galactic center is invisible to us, Sagittarius is enormously rich in visible and interesting objects. Indeed it is worth spending several hours—spread over several nights, of course—slowly scanning the region with your binoculars and seeing how many of them you can pick out; you should be able to spot close to half a dozen open clusters and even more globular clusters, though some will be hard to distinguish because they lie amid such dense star fields. The rich open cluster M-2, indeed, is hard to distinguish even in a small telescope because it lies at the edge of a star cloud which looks like—though it is not—an even brighter cluster.

The three most conspicuous objects in Sagittarius are not classified as clusters at all, but as "diffuse nebulae"—that is, clouds of glowing gas illuminated by nearby stars.* I will now tell you a secret you won't find in other popular astronomy books—or at least none that I've looked at: To the naked eye, through binoculars, and sometimes even through a small telescope, *a diffuse nebula looks very much like an open cluster.* There are two simple reasons for this. First, all diffuse nebulae are associated with clusters of stars—in fact it is only the reflected light of these stars that makes visible the nebulous clouds of dust and gas that give them their name; lacking these, they would be what are called "dark nebulae."† Second, diffuse nebulae *are* diffuse; their light is generally so dim as to show up only in a sizable telescope. Recently, in a popular astronomy magazine, I noted the casual statement, apropos of two of the three conspicuous nebulae in Sagittarius, that *"on nights of good seeing,* an *eight-inch* telescope will bring out much intricate and delicate detail . . ."* (my em-

---

*The "diffuse" distinguishes them from so-called "planetary nebulae," which will be discussed—briefly—in the section on telescopic viewing; see page 109.

†By the same token, most open clusters are also associated with nebulosity, though long-exposure photographs are usually necessary to see it; a case in point is one of the best-known open clusters, the Pleiades.

phasis). The statement, in other words, is probably irrelevant to at least ninety-five percent of amateur stargazers, who must make do with smaller telescopes or none.

The two nebulae in question, M-20 and M-8, lie respectively about 7°/8° degrees almost due north of Gamma-Sgr, which marks the star marking the tip of the right-hand "arrow" in this constellation. (That is, they will lie more or less directly above the star, depending on how close the latter lies to due south.) You should have no difficulty in picking them out—in fact on any clear night you can see them with the naked eye. Through binoculars they will look like open clusters, albeit rather more spread out than most of those you have thus far looked at; on a really clear, dark night, you *may* perceive a dim sort of glow surrounding the clustered stars, which is of course the nebula proper. What you will *not* see is the structure of these nebulae—the dark lanes which divide M-20 into the three parts which give it its name, the Trifid Nebula, or the other structural details that characterize the even brighter M-8, the Lagoon Nebula. Nor will you have any better luck with the Horseshoe or Omega (Ω) Nebula, M-17, which lies about 7 degrees north and east (*i.e.*, above and to the left) of M-20—not even to note that the name, as is so often the case, has not much relationship to reality, since the nebula's shape has been described as an "L" and a "2" as well as a horseshoe. If and when you graduate to a telescope, you will be able to see these nebulae as nebulae; meanwhile, enjoy them for what they are to the eye: unusually bright open clusters.

Less than a degree from M-20, above it and to the left, lies a "real" open cluster, M-21; indeed the two objects are so close that you will be doing well if you can figure out where one leaves off and the other begins. About 4° above M-20, and a little to the right, you should have no trouble spotting another open cluster, M-23, while two others, M-18 and M-24, lie within a degree or two of the Horseshoe Nebula.

Globular clusters are no less profuse; the most conspicuous is M-22, which forms an almost perfect right triangle with Phi-Sgr, marking the tip of the left-hand "arrow," and Kaus Borealis, marking the top of the right-hand "bow." Several other clusters lie to the right of M-22, while another group occurs in the space between the stars marking the bottoms of the two bows. But these, being only seventh and eighth magnitude, are less easily seen, especially the latter group which in most latitudes will lie uncomfortably close to the southern horizon.

After the celestial riches of Sagittarius and Scorpius, almost any other part of the sky can seem an anticlimax. Nonetheless, the late-summer (or late-evening-in-early-summer) constellations of the eastern and northeastern sky are by no means devoid of interest, and one of them contains an altogether unique object whose like cannot be seen elsewhere in the heavens except with the aid of a sizable telescope.

The constellation in question is Andromeda, which (as noted earlier) can be most easily located by following an imaginary line from Polaris through Caph, on the upper right of the

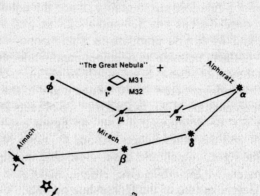

ANDROMEDA

The symbols used on this map are identified in Appendix A.

"W" in Cassiopeia. This will lead you to second-magnitude Alpheratz, lying at the left-hand corner of the Great Square. Now fix your attention on the two curved lines of stars stretching left from Alpheratz, specifically the two stars in the middle of these lines—*i.e.*, second to the left, beginning with Alpheratz. Run your eye upward from Mirach, the lower and brighter of the two, to the upper, and continue on the same line for about the same distance. On a very clear, dark night, if your eyes are good, you may be able to see a faint hazy patch; your binoculars will show it clearly as a hazy, oval glow. This is the Great Nebula in Andromeda. Despite its name, it is not a nebula but a galaxy—one of the enormous "island universes" scattered by the thousands through space. Like our own galaxy, the Great Nebula (M-31) contains billions of stars; estimates run as high as 200 billion, making it considerably larger than our own galaxy, with perhaps 100 billion. What we see, in binoculars or an amateur's telescope, is in fact only part of that number. Time exposure photographs through observatory telescopes reveal that the entire galaxy stretches over more than 3° of sky; were our eyes sufficiently sensitive to see it whole, it would be bigger than the moon. The elliptical glow in our binoculars is merely the denser, central portion of the galaxy, in which, as in our own, the stars lie so close as almost to jostle one another.

The fact that this dense mass of 100 billion or more stars glows so faintly, even in binoculars, obviously means that it is very far away. In fact, it lies something like 1.5 million light-years from earth—a figure that is meaningless to almost anybody but an astronomer. A light-year is, of course, the distance light can travel in a year, but most of us find it hard to grasp the notion of light "traveling" at all, since in our everyday experiences it moves instantaneously—"quick as a flash," in fact. Translating the figure into miles is no help; a light-year is close to 6 trillion miles, meaning that M-31 lies some 10^{19}—one followed by 19 zeroes—miles away. But let us see if we can get some dim notion of what this distance means.

Light, to begin with, travels about 186,000 miles a second.

The moon's light takes almost 1½ seconds to reach us; if you were talking by radiophone with an astronaut on the moon and asked him a question, it would take some three seconds for the question to reach him and his answer to get back to you.

The sun is, of course, much farther off than the moon; its light requires some 8⅓ *minutes* to reach us. Moving out to the borders of the solar system, to Pluto, most distant of the sun's planets, we can calculate that its light must travel for some 8 hours and 40 minutes before it reaches the large telescopes needed to see this far-off body.

Beyond the solar system, the distances begin to stretch out. Rigil Kentaurus, a Southern Hemisphere star even brighter than Arcturus, is more than 4 light-years away, and it is the closest star to us. The light from Arcturus, when I looked at it last night, had been on its way since 1943—the year the Germans were defeated by the Russians at Stalingrad and kicked out of North Africa by the Anglo-Americans; the reddish rays of Antares had been traveling since about 1750 —around the time Ben Franklin was flying his famous kite. Deneb, one of the more distant among the brighter stars, lies some 650 light-years away, meaning that its light began traveling more than a century and a half before Columbus landed in the West Indies.

Moving out still farther, the Lagoon Nebula in Sagittarius is so far off that its light, showy in binoculars, began moving toward us some five centuries before the birth of Christ—the period when the Greeks and the Persians were squaring off for their historic war. The great globular cluster M-13 in Hercules is far more distant; the light of its million or so stars started on its way some 20,000 years ago, about the time the Cro-Magnon artists were scratching their pictures of mammoth and bison on the walls of their caves.

But the ghostly glow of M-31 has been traveling for far, far longer than that. When its light began its journey, the Cro-Magnons were well over a million years in the future. Their ancestors (and ours) were represented by scattered bands of hairy creatures trotting about the African veldt, wielding clubs made from torn-off tree branches and butchering their game with pebbles they had battered into the crude semblance of tools.

A million and a half light-years is that long.

Of the many thousands of galaxies that astronomers have tracked down with their giant telescopes and catalogued, M-31 is by all odds the nearest visible in our latitudes. (Two smaller galaxies, the Lesser and Greater Magellanic Clouds, are conspicuous naked-eye objects in latitudes below the equator; they are much closer than M-31, and in fact are thought to be satellites of our own galaxy.) M-31 is the only Northern Hemisphere galaxy visible to the naked eye, and the only one that "looks like anything" in binoculars or even a small telescope. The brightest of the others are only about eighth magnitude (M-31 is fourth magnitude), meaning that even in a medium-sized scope they appear as tiny blurs of light, identifiable for what they are only if you know exactly where your telescope is pointed.

To the eye, M-31 is not a very impressive object. To the imagination, however, it is staggering. To gaze at its shimmering, spectral radiance, to reflect that it contains more stars than the Milky Way, and that among these stars there may be millions of planets not unlike earth, populated, it may be, by beings as intelligent as we are—or more so—is to me, at least, one of the most extraordinary experiences that any star-watcher can have.

Moving north (*i.e.*, left) from Andromeda brings us to the conspicuous "W" of Cassiopeia. Since it lies in the Milky Way, its star fields are rich and include many open clusters —which (as with the clusters in Cygnus and Sagittarius) you may have trouble distinguishing from their starry surroundings. Especially rich is the region around Delta-Cas, the star marking the left-hand "V" of the "W"; a careful sweep of the area with binoculars should reveal at least two or three clusters, and sharp-eyed observers will spot even more.

The brightest clusters in this region, however, do not lie in Cassiopeia but in its neighbor, Perseus; they are the famous Double Cluster. They can be found most easily by starting with Mirfak, brightest star in Perseus and set in an unusually brilliant star field which makes the star almost unmistakable once you have seen it through binoculars. Moving upward about 4°, you reach Gamma-Per, a fairly conspicuous third-magnitude star. If you now sweep toward Delta-Cas, you will find the double cluster conspicuous in your binoculars about halfway between it and Gamma-Per; on clear nights, you should be able to see it with the naked eye.

Another interesting object—and for the amateur, a unique one—in Perseus is the star Algol, second brightest in the constellation. It can be found easily enough by relating it to Mirfak and Almach, the second-magnitude star lying at the end of the lower and brighter line of stars stretching to the left from Alpheratz in the Great Square: the three stars form a right triangle, with Algol lying at the right angle. Algol is a variable star, and one, moreover, whose variations in brightness can—with luck—be observed with the naked eye. Every 2.7 days—

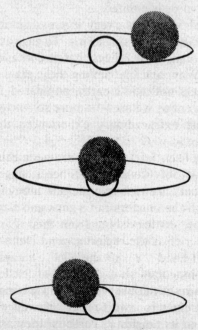

The star Algol shows regular, marked and relatively brief dips in brightness. These occur when the dimmer member of this binary pair passes in front of, and partially eclipses, the other.

about 2 days and 17 hours—its light wanes from second magnitude almost to fourth magnitude—that is, from virtually as bright as its neighbor, Mirfak, to markedly dimmer. The process is very rapid as such things go: the star dims for about 4½ hours, remains at its minimum for about 20 minutes, and then returns to normal over the next 4½ hours, with the next cycle beginning about 2 days and 8 hours later.

These sudden, intermittent and pronounced changes in brightness make Algol something of a rarity among variable stars, most of which brighten and dim in a more or less smooth and continuous manner. The astronomers have deduced that Algol is not, technically, a variable star at all, but a double star, one of which regularly passes in front of, *and partially eclipses,* the other. The variation, incidentally, is considerably more pronounced than that of most variables, most of which change by half a magnitude or less. Polaris, for example, is a variable with a period of about 4 days, but you are very unlikely to notice this, since the range in brightness is only a little over a tenth of a magnitude—from 2.48 to 2.62.

The name Algol, like most star names, comes from the Arabic, and means "the demon" (our word "ghoul" comes from the same source). To the Arabs, Algol's regular drops in brightness meant that the demon was blinking his eye. If you want to see this, the simplest way is routinely to check Algol every chance you get, comparing it with the two stars, Mirfak and Almach, used to locate it. Most of the time it will be almost indistinguishable from Mirfak in brightness. If, however, it is noticeably dimmer, you have caught it at or near one of its eclipses; at its minimum, it will be dimmer than Almach. Now wait for a couple of hours and check it out again; this should give you some notion of whether you have caught the star on the upswing or downswing. As a further check, you can, with luck, observe it again exactly eight days (almost three complete cycles) later; if it appears noticeably brighter than it did the first time, it is on the downswing, and will continue to get dimmer for two to four hours; if dimmer, it is on the upswing and will regain its full brightness some four hours later.

By late September—or even late August, if you are willing to wait until midnight—some of the interesting fall and winter constellations will be visible toward the east. One of these, Auriga, is easily identified by its principal star, Capella, which is almost exactly as bright as Vega; if the horizon is clear and your timing is right, you may be able to view the brilliant trio, Arcturus, Vega, and Capella, spread across the sky from west to east. Four other conspicuous stars, of second and third magnitude, form with Capella an irregular pentagon which makes Auriga virtually unmistakable, once you know where to look for it—below Perseus, and to the left. Capella is the most northerly of all first-magnitude or brighter stars.

The real treat among the autumn constellations, however, is Taurus, a constellation for which I have a certain mild affection because it is my birth sign. I remember one late summer

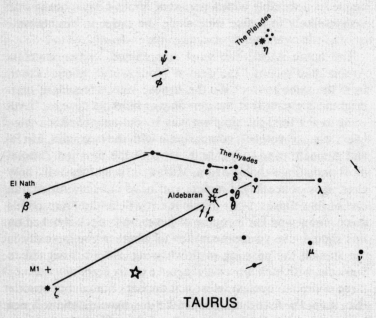

TAURUS

The symbols used on this map are identified in Appendix A.

night when I was just beginning to find my way around the heavens (I didn't yet have even a small telescope) seeing low on the eastern horizon what seemed to be a dim cloud of stars; through the binoculars they emerged as a conspicuous, jewel-like cluster. I had never seen them before, but somehow knew that I was looking at the Pleiades, known to the ancients as the harbinger of autumn. To the naked eye, the Pleiades are conspicuous but not especially attractive—though you can test your eyes by seeing how many of its brighter stars you can pick out with the naked eye (five is a fair average). Nor, for that matter, are they much more interesting through a tele-scope, since you can look at only part of the group at any giv-en moment. But through binoculars they form a cluster of dia-monds that millionaires would bid to possess. The principal stars are arranged in a tiny dipper, and scattered around and among them are over 100 other stars—though binoculars will not show all of these. If you can stargaze at any time in late summer, make sure you see the Pleiades—they're worth wait-ing for.

Taurus also contains another, almost equally famous clus-ter, the Hyades. This group of stars is less conspicuous than the Pleiades because it is more dispersed; its stars are scat-tered in the form of a rough "V" lying sideways, with orange-red Aldebaran, the first-magnitude star that dominates Taurus, lying at the tip of the lower arm (it is not itself part of the cluster). In the imaginations of those who named Taurus, the Hyades were thought to indicate the face of the bull (only its head and shoulders appear in the sky), with Aldebaran marking the base of his lower horn (or perhaps one of his eyes). The tip of the other horn (both are long) is marked by the bright second-magnitude star El Nath, which visually is part of the Auriga pentagon, but technically is assigned to Taurus.

There is one other group of stars which, though not strictly a late summer or early fall constellation, should definitely not be missed. This is Orion, which, dominating the southern heavens for most of the winter months, supplies the most

ORION

The symbols used on this map are identified in Appendix A.

spectacular sky show of any constellation—and is, incidentally, the most easily recognized, possibly excepting the Big Dipper.

In late September, Orion will be well up in the eastern heavens by about 1:30 A.M. or so, but in late August it can only be
seen by staying up until shortly before dawn—or, more likely,
getting up before dawn (say, four A.M.). Its chief stars are arranged in a long, irregular quadrilateral, lying at this season
more or less on its side. At the upper right corner is blue-white
Rigel, almost as bright as Vega (which will be still visible low
in the west), while at the opposite, lower left corner is reddish

Betelgeuse, the most brilliant variable star in the sky. At its brightest, it is noticeably brighter than Altair, and even at its dimmest—about 2 years 10 months after its peak—it is still almost exactly as bright as Deneb. Above Betelgeuse is Bellatrix, a bright second-magnitude star, even bluer than Rigel, while below Rigel, at the lower right, is yet another blue-white star, somewhat dimmer than Bellatrix, which is second-magnitude Saiph. These four stars are supposed to mark the shoulders (Betelgeuse and Bellatrix) and knees (Saiph and Rigel) of the great Hunter. Across the middle of the quadrilateral, lying more or less vertically at this season, are the three stars of the hunter's "belt"—from bottom to top, Alnitak, Alnilam (both second magnitude) and Mintaka, whose slightly variable light fluctuates around the borderline between second and third magnitude. (As a matter of interest, Mintaka lies almost precisely on the celestial equator—the imaginary line in the heavens lying directly above the earth's equator—meaning, of course, that in northern latitudes it lies not overhead but toward the south.)

To the right of the belt, finally, is the "dagger"—a close line of three stars—the middle one of which, however, looks somewhat hazy. Binoculars show that the haze is due to a cloud of luminous gas surrounding the "star" (actually, as a telescope shows, it is four stars)—the Great Nebula in Orion, or M-43. Unlike the other Great Nebula—M-31 in Andromeda, which as we noted is really a galaxy—Orion's nebula is a true nebula, and the only object of that class that can easily be seen as such in binoculars.

Enumerating the stars in Orion, however, can give only a dim idea of its beauty; you have got to see it. If you are willing to stay up a little longer, you may, if it is late enough in summer, see a special "added attraction": the star Sirius, whose magnitude of -1.42 makes it easily the brightest star in the sky, and, except for the sun, moon and the brighter planets, the brightest object in the heavens. Sirius can be located by following the line of Orion's belt downward*—but once seen, it is unmistakable, a magnificent sky beacon which represents a fitting period to this section of the book.

*Following the belt upward leads you to Aldebaran.

Chapter 4

ABOUT TELESCOPES
How to Buy and Use Them

BY this time, you should be well aware of how much difference even the simple and inexpensive optical aid of binoculars makes to what you can see in the sky. A telescope makes far more difference. Roughly speaking, what you can see with even a 3-inch telescope compares with what you can see with binoculars as the latter compares with the naked eye view; a 6-inch scope will add as much again. As I have already suggested—and as the balance of this chapter will make clear—learning to use a scope is a much more complicated business than using binoculars, but it's worth it.

Before learning to use a scope, however, you must first buy one, and to do this intelligently you will need to know something about what you're shopping for.

Telescopes come in two basic types, "refractors" and "reflectors"; the difference is in the method employed to concentrate light. In a refractor, as in binoculars, light rays are concentrated by passing through, and being bent (refracted) by, a lens; in a reflector, the same effect is obtained by reflecting the rays from a curved mirror. In both types, as in binoculars, the concentrated image is then magnified by eyepiece lenses, but whereas most binoculars have a single, fixed eyepiece—and therefore a fixed degree of magnification—telescopes typically come with several different eyepieces that can be substituted for one another in order to vary the magnification.

Several makes of telescope have objectives (the light-gathering element) which *combine* mirror and lens, principally in order to make the telescope tube shorter, and the instrument more compact and portable. Without exception, however, they are terribly costly—the cheapest sells for over $600—and should therefore not even be considered except by people who are really "into" astronomy—or well into the upper-income brackets.

People with little or no knowledge of telescopes are prone to ask, of a particular instrument, "How much will it magnify?"; I have lost track of how many times this question has been put to me by curious friends. This misconception is fostered by some manufacturers, whose advertising boasts that their instrument "enlarges 300×" without mentioning that it will almost certainly *not* give a clear image at that enlargement—and that in any case the user will seldom have any reason to enlarge that much, and many reasons not to.

A telescope's maximum degree of enlargement gives the power at which the image starts to blur, like an over-enlarged photograph in which details dissolve into a grainy texture. It is determined by the instrument's "aperture"—the diameter of its main lens or mirror—and amounts to roughly 100× per inch of aperture. Thus my 6-inch refractor can, theoretically, enlarge its image up to 600×. However, this would be only under absolutely ideal conditions: brilliantly clear weather, plus a minimum of atmospheric turbulence—the phenomenon that makes stars twinkle. As a practical matter, the maximum *useful* enlargement is more like 50× per inch, and in many situations your image will start to blur at half that figure.

Quite apart from this limitation, there are excellent reasons why working at high magnifications is undesirable—notably, our old friend "shake." In comparison with hand-held binoculars, of course, a telescope on a fixed mounting has very little shake indeed, but a good part of the difference is cancelled out by the fact that the degree of enlargement is much greater—and the shake is enlarged along with the image. At even 100× enlargement, a vibration of the scope of only 1/500 inch—about the diameter of a human hair—will produce an apparent movement of nearly ¼ inch in the image, and that much vibration can be produced by no more than a lively puff of wind!

No matter how much you enlarge the image in your telescope, you won't make it any brighter—and in some cases will make it dimmer.* Increased enlargement can clarify and separate details that were previously more-or-less melded together—but it won't put in any details that weren't there to begin with.

The amount of detail you can see is limited by the instrument's "resolving power," and this, too, depends on its aperture. Resolving power, measured in seconds of arc ("), expresses the angular distance between two objects required for the scope to show them as separate and distinct, rather than fused together into a single image. When you look at the famous "double double" in Lyra (see page 103), for example, your eyes—unless they are exceptionally keen—see it as a single star; binoculars will resolve it into two stars, while a 3-inch telescope at high magnification will, if the seeing conditions are good, further resolve it into four.†

The theoretical resolving power of any scope (or binoculars, for that matter) can be determined by a simple formula: 4.5 divided by the aperture in inches, or 112 divided by the aperture in millimeters. Thus my 6-inch reflector will theoretically resolve down to less than 1″ (4.5/6) of arc, while my 60-millimeter refractor has a lower limit just under 2″ (112/60). As with enlargement, however, resolution at the mathematical limits requires optimum conditions and optimum eyesight; if you can consistently resolve objects at twice your theoretical limit, you will be doing very well.

---

*This is the case with the moon, the planets, and gaseous nebulae, in which the image is an area rather than a point, as are star images. The more the area is enlarged, the more the light from it must be "spread out," like a pat of butter over progressively larger and larger slices of bread. The *total* light which the image emits is unchanged—but the image itself looks dimmer. With stars, whose images remain points no matter how much enlarged, this does not occur.

†This is what the books say; in practice, my 6-inch will just about do it.

Yet another consideration in selecting a telescope is the fact that its light-gathering power increases very rapidly with increasing aperture—specifically, as the *square* of the aperture; thus a 6-inch scope, with twice the aperture of a 3-inch instrument, has *four times* the light-gathering power.

What all this adds up to is that you should get the biggest aperture you can afford—so long as it is suitable for the kind of observations you plan to make and (especially) the kind(s) of place(s) you expect to make them. What it also adds up to is that for most purposes you will be better off with a reflector than with a refractor, because the former gives you much more aperture for your money. The reason is quite simple: a major part of the manufacturing cost of any scope comes from grinding its objective. A refractor objective, for optical reasons, consists of two lenses cemented together, which between them have *four* surfaces, each of which must be ground to a high-precision curve if the lens is to yield a clear and undistorted image. In addition, since the light passes *through* the lenses, they must be made of glass containing no bubbles or other irregularities. A reflector mirror, by contrast, has only *one* curved surface that must be precision ground—nor do irregularities in the interior of the glass make any difference, since the light never gets to the interior.* The difference in cost is suggested by the fact that at this writing one can buy a *6-inch* reflector, complete with mounting and all the basic accessories, for about $300—which is considerably less than the cost of a *3-inch* refractor.

With such differences in cost, why would anybody—or at least anybody knowing anything about telescopes—buy the refractor at all? The main reason is portability. My 6-inch reflector weighs in excess of fifty pounds; even moving it 30 feet along the deck of my summer home is a bit of a chore. My 2.4-inch refractor, by contrast, weighs only about 20 pounds and can easily be set up almost anywhere; indeed if I wanted to observe at some fairly distant spot not too far from a road I could fit it into the trunk of my car. A 3-inch reflector, of

*Some modern observatory telescopes, in fact, have mirrors made of the completely opaque ceramic substance called cermet.

course, would be no less portable—and would have cost me a good deal less. As a practical matter, however, most manufacturers don't make reflectors this size (perhaps in part for the same reason the American automobile industry much prefers to make large cars: there's more profit in them), and some don't even make a 4½-inch reflector, which would be almost as portable as a 3-inch refractor.

Let's now sum up the considerations you should have in mind in deciding on a telescope. If you can observe from a more-or-less "permanent" site—one where you can safely leave your scope (covered up, of course) set up for days or weeks at a time (the deck of a summer cottage, or even an open stretch of ground nearby)—then by all means get a reflector—preferably a 6-inch job; a 4½-inch or 3-inch scope (if you can get one) won't save you all that much money, and will show you a great deal less in the sky.*

If, on the other hand, you must do your sky-watching on a catch-as-catch-can basis, setting up your scope in relatively inaccessible places (a beach, a distant hilltop) and taking it down each time, you will be better off with a smaller and lighter model—a 3-inch or 4½-inch reflector if you can get one, a 3-inch or 2.4-inch refractor if you can't. Check the instrument's weight against your own carrying capacity and the distance you'll need to carry it.

Finally, if portability is important but price isn't (you lucky devil, you!) get one of the "compound," catadioptic (reflector-refractor) scopes such as Criterion, Celestron or Questar—again, in the largest aperture you can afford.

Having decided on the type and size of scope you want, you must next consider the accessories you will need with it: a mounting, and eyepieces. The mounting consists of an arrangement of shafts or pivots permitting the scope to be turned in any direction; it is supported by either a tripod or a pedestal—a large iron pipe with three "feet" at the bottom. The pedestal is steadier than the tripod, but also heavier, if portability is a consideration. It is standard on most 6-inch (or

---

*It will also be much easier for somebody to pick up and walk off with, if that's a consideration.

larger) scopes and some 4½-inch ones, while the others almost always employ tripods.

The mounting proper, which actually holds the scope and permits its adjustment, can be of two possible types: *altazimuth* and *equatorial.* Without going into the differences between them in detail (these are explained on page 89), the altazimuth is much cheaper—it is, in fact, sold only with the smallest and cheapest "beginner's" scopes—but in every other respect is much inferior to the equatorial, which you should therefore get if at all possible.*

An equatorial mounting may or may not include a "clock drive"—an electric motor that will keep the scope centered on a given object for as long as you care to look at it. The motion of the stars across the sky, when you are observing them at telescopic magnification, is sufficient to carry them out of the instrument's field of vision in a minute or less (the higher the magnification, of course, the quicker). To keep their image in view, therefore, the scope must be moved either by motor or by hand.

Telescope manufacturers prefer to sell you a clock-drive mounting, for the same reason car dealers prefer to sell cars with as many "options" as possible: They make more money. In fact, the clock drive possesses both advantages and disadvantages, depending on how and where you propose to use your scope.

First the disadvantages. An electric drive obviously requires electric power—which means that you will not be able to use it more than (say) 50 feet from an electric outlet—and even this will require a long extension cord.† Moreover, the power requirement complicates setting up the scope: you will need to connect your extension cord each time you use it, then disconnect and coil it when you've finished your observations (rain, or even dew, in the socket of the cord will short it out next time you turn on the current).

*The only exception is if you are getting a scope for a child who may lose interest in the whole business in a month or two; here, obviously, it makes more sense to get the cheapest scope you can find and "trade up" if the kid's interest holds up.

†Some clock drives, however, can be run off a 12-volt car battery, which may somewhat simplify the power problem in locations far from a house; check this with the manufacturer.

If the prospective site(s) of your scope make electric drive feasible at all, you will need to weigh the disadvantage in setting up against the advantage of the drive, which is essentially that—assuming your scope is properly set up and adjusted—it will keep an object in view with no further adjustment, for as long as you want. Just how important is this?

If you are a camera freak and plan on using your scope for celestial photography in the future, the electric drive is absolutely essential; with trivial exceptions, telescope sky photography requires time exposures, which in turn require that the scope be held absolutely steady on target—and this is quite impossible to do by hand.*

Electric drive is a convenience—though not a necessity—for work at magnifications much above 100×; the higher the magnification, the greater the convenience. At (say) 200× magnification, your target star will move out of the field in thirty seconds or less. This means that, if you are changing eyepieces to increase the magnification, you may lose the star entirely by the time the high-power eyepiece is in place and focused. On the other hand, you are unlikely—for reasons I have already mentioned—to want to do much (if any) work at such magnifications; indeed, unless your scope is at least 6-inch, 200× will seldom be worth bothering with, if at all.

In general, I would sum up the matter like this: assuming you have convenient access to power, get the electric drive *provided you can also use hand controls if desired:* technically, this is called "slow motion manual control on the polar axis." *Make sure your scope has this in any case;* it is a major convenience even with electric drive, and an absolute necessity without it. If the scope cannot be controlled *both* electrically and manually, stick with manual controls, or (better) look for another manufacturer.

Some scopes also have "slow motion manual control on the equatorial axis." This is not a necessity, but is a major con-

---

*Note, however, that according to a Consumers Union study of inexpensive scopes, few, if any, of their *clock drives* will track with sufficient accuracy to make photography feasible except at fairly short exposures. If there is a serious chance that you will want to do sky photography, the best plan would be to consult some more experienced amateur or to write to one of the astronomy magazines, for advice on equipment.

venience, for reasons explained later on. Unfortunately, some manufacturers have eliminated this feature over the years, I assume in an effort to keep prices down. My own feeling is that this is false economy; I would rather pay an extra $25 for the convenience of equatorial slow motion—or at any rate, have the option of doing so. Check out whether the scope you are thinking about buying includes this feature, and if not, whether you can get it as an option.

Finally, we come to the matter of eyepieces. Virtually all scopes are sold with two or three eyepieces of different powers; some manufacturers even give you a selection from which you can choose. To understand what you are getting (or choosing), however, we must get a bit technical.

Eyepieces are conventionally classified, not by magnification, but by focal length, which ranges from 50 mm down to 6 mm or less—the shorter the focal length, the higher the magnification. The reason for this classification is that the magnification of a given eyepiece doesn't depend simply on its *own* focal length but also on the focal length of the objective. Knowing this last figure (which should be part of the telescope's specifications), you can translate eyepiece focal length into magnification by dividing it into objective focal length; the formula is Magnification $= F_o/F_e$. To take a couple of examples: my small refractor has an objective focal length of 900 mm; used with my 23-mm eyepiece, the magnification is 900/23 or about 40×; the 6-mm eyepiece gives me 900/6 or 150×. My larger instrument has a focal length of 1250 mm; eyepieces of 50 mm, 18 mm, 12.7 mm and 7 mm* give me magnifications of 25×, about 70×, almost 100× (98-plus if you're fussy) and almost 180×.

In order to understand the following recommendations, you will need to translate the focal length of the eyepieces you are considering into magnification, *given the focal length of the telescope you are considering.* Broadly speaking, three eyepieces that range from 70× to 180× will give you high enough

---

*The reason for some of the odd sizes is something that perhaps the manufacturer could explain; I can't.

magnification for almost any purpose, and low enough for most purposes.* If you can, add a low-power, wide-field eyepiece—say, 30× or 40×. This is useful for examining relatively large objects—including the moon and the nearer open clusters—and also, because of its wide field, for "scanning" across rich sky regions (*e.g.*, in Sagittarius and Scorpius) for interesting objects, which can then be examined in close-up by changing to a more powerful eyepiece. Low-power eyepieces, however, are not standard equipment on most scopes except the smallest, so you'll have to pay extra. An important point is that you should avoid low-power eyepieces with a focal length of more than 40 mm. The reason is that in most 50-mm eyepieces, such as I have, the size of the lens is somewhat bigger than the pupil of the eye, the result being that you cannot see the entire field at once—thereby losing much of the original "point" of the low-power attachment.

Becoming increasingly common is the "zoom" eyepiece, which, as its name indicates, allows you to change magnification over a considerable range without changing eyepieces. Zooms are standard on a few scopes, and can be gotten as options on most others. Their advantages, for many purposes, are considerable, but it is probably best to let their purchase—they are fairly costly—wait until you are sure you want to spend the money.

Another piece of (usually) optional equipment is the "Barlow lens," which in effect doubles or triples the power of the eyepiece it is used with. Don't even consider this until you have already used your scope for a while, and have decided whether such a gain in magnification will be worth the cost of the Barlow.

One other feature about your prospective scope that you ought to check out is the "finder scope." This is a small scope, mounted on the main one, which as its name implies aids you in finding objects. The main scope necessarily has a very narrow field of vision—even at low powers, not much more than a degree—so that it does not show the larger sky

*This, of course, is why I picked these three eyepieces for my own use.

patterns you have learned to rely on to find your way about the heavens. The finder scope magnifies only feebly—usually 6× or 7×—but in exchange, gives you a field approximating that of binoculars, and thus large enough so that you can recognize sky marks. Like a telescopic rifle sight, a finder scope comes equipped with "cross hairs," which mark the exact center of its field.

The larger the finder scope, the better, of course; the wider its aperture, the wider its field and the more clearly it will show dim objects. Another consideration is that it should preferably show a field at least as large as your binoculars. Assuming these are the usual 7 × 35s, a 7 × 35 finder would be desirable if you can get it; if not, a 6 × 30 finder will show the same size field, though less brightly, while an 8 × 40 will also show the same field, but more brightly. A 6 × 24 finder, such as I have on my small telescope, should be avoided if at all possible, since its field is both narrow and dim; a 7 × 50 or 8 × 50 finder, on the other hand, has a wider-than-average field, but the gain in brightness from the larger aperture more than makes up for any possible confusion between the finder and the (smaller) binocular fields.

The finder, whatever its size, is mounted on the main tube in an arrangement of two rings, each set with three screws that hold the finder firmly in position. A few cheap scopes have only a single ring in the mounting; these should be avoided, as the finder gets out of adjustment much too easily.

Setting Up Your Telescope

How (and of course where) you set up your scope will have a lot to do with how much you can see, and how easily. An accurate setup will align your scope with the celestial coordinates of "latitude" (declination) and "longitude" (right ascension) thereby enabling you to find hard-to-spot objects more easily and, having found them, keep them (or any other object) in view *much* more easily. Only with an equatorial mounting can you achieve such an alignment—which is, of course, the reason for getting it to begin with.

The very simple and cheap altazimuth mounting needs no setup to speak of. It embodies just two pivots—a vertical one, on which the scope can be turned in any compass direction (azimuth), and a horizontal one, on which the elevation (altitude) of the scope can be varied from 0° (the horizon) to 90° (the zenith). If that's the kind of mounting you have, you can skip directly to the bottom of page 92.

An equatorial mounting also has an azimuth pivot and an altitude pivot, but the former is normally adjusted once at the beginning of the evening (if you have a "temporary" site) or at the beginning of the season (for a "permanent" site), and perhaps not even that often. The altitude pivot is normally adjusted only once, when you first set up, and then permanently locked. In other words, during the normal course of observation, the scope is *not* free to move on either its azimuth or its altitude pivot.

To enable it to be pointed at any part of the sky, an equatorially mounted scope can be turned on two other pivots or steel shafts, known as the polar (or right ascension) axis and the equatorial (or declination) axis. The polar axis, when properly set up, points at the celestial North Pole (*i.e.,* almost exactly at Polaris), while the equatorial axis, as you have probably guessed, will point to some point on the celestial equator. By turning the scope on the equatorial axis, you can adjust the declination to anything from +90° (the approximate declination of Polaris) to as far south as the horizon will let you see. (In the mid-latitudes, the southern horizon will lie somewhere between declination (dec) −35° and dec −60°, though because of haze you will be doing well if you can examine objects at dec −25° and dec −50° respectively.) Similarly, turning the scope on its *polar* axis allows you to point it at any right ascension from 0 hrs 01 min to 24 hrs 00 min. In addition, by using the slow motion control on this axis, you can keep the scope zeroed in on its target as the latter revolves around the celestial pole, which it does, of course, at the rate of one complete revolution every 23 hrs and 56 min (remember that stars rise 4 minutes earlier per day).

But the mounting will only work this way if it is properly set

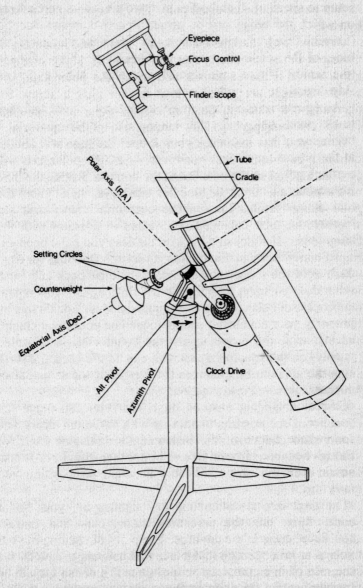

A 6-inch reflecting telescope, of the general type shown here, is an excellent instrument for the beginner—unless portability is important.

up—specifically, if the telescope is level, and if the polar axis in fact points to the celestial pole. This is how you do it, step by step.

To begin with, assemble the scope, finder and mounting as indicated in the directions that accompany it. Make sure you know which is the equatorial and which the polar axis, and which controls are which—slow motion plus a screw for clamping the scope in position on the polar axis, another clamp screw and perhaps slow motion also on the equatorial.

When you first assemble your scope, the tube will almost certainly be more or less horizontal—*i.e.*, the polar axis will be pointing, not at the pole but at the horizon. Rotate the tube on the polar axis until the tube lies sideways, then tighten the polar clamp firmly. Loosen the equatorial clamp and see whether the tube balances—*i.e.*, stays in position with the clamp loose. If it doesn't—it won't unless you have been extraordinarily lucky in assembling it—loosen the rings (or ring) which hold the scope tube and slide the tube back and forth until it does balance, then tighten the rings again. Now repeat the process of balancing on the equatorial axis. You do this by tightening your equatorial clamp, loosening your polar clamp, and balancing the scope against the counterweight mounted (usually) on the equatorial axis; this can be slid back and forth along the axis until it balances the weight of the scope, after which it is screwed into position.

Careful balancing ensures, first, that you can move the scope from one position to another with minimum effort, and a minimum chance of the instrument's damaging itself (or you) by whipping around in one or another direction. It also ensures that the scope will stay in whatever position you move it to.

The next step is collimating or "sighting-in" your finder scope. This, like the preceding steps, can—and in fact should—be done in the daytime. The point of sighting-in your finder is to make certain that it is in fact pointing in exactly the same direction as the main scope, since if it doesn't it will be virtually useless for pointing the scope at a target. In this respect, as in others, it resembles a telescopic sight, which must also be sighted-in if it is to give the rifleman any help. In spy movies and TV, the assassin simply snaps his telescopic sight into position and is ready to open fire—but not in real life, since the sight's adjustment is crucial to accurate aiming.

You sight-in your finder by picking some small but conspicuous object a considerable distance away—at various times I have used a transformer "pot" atop an electric pole perhaps a mile off, or the top of the Pilgrim Monument in Provincetown some seven miles away. Set your finder into its rings, tightening the screws on one ring fairly firmly, and as evenly as you can manage by eye, but leaving the screws of the other ring almost loose. Now move the tube—if you've balanced it properly, this should be easy—until your sighting target is exactly in the cross hairs of your finder. You will find that the image looks upside down—which, as we shall see later, is almost but not exactly true. Next, using your lowest-power eyepiece, see if the target is also visible in the main scope; you will need to focus the eyepiece to get a clear image. If it isn't visible, move the tube back and forth slightly in various directions—use slow motion if you have it—until you can see the target in the exact center of your field. Tighten both the polar and the equatorial clamps firmly.

Now look again through your finder scope; if (as is likely) the target has moved off center, adjust the three screws on the "loose" ring until it is once again exactly in the cross hairs. Check to make sure you can still see the target in the main scope, then, bit by bit, tighten the three screws on the "tight" ring, moving from one to the next and tightening a little more each time, until all three are as tight as your fingers will make them. Again check that both the finder and main scope are on target, and do the same with the "loose" ring, tightening each of the three screws more or less to move the target into its exact position against the cross hairs.

When you have finished this process, all six screws, on the two rings, should be as tight as you can make them, the target should be right in the finder's cross hairs, and should also be visible in the main scope even through a fairly high-powered eyepiece. Take as long as you need to get this exactly right;

the time will be saved many times over when you begin targeting in, not on an obvious and conspicuous daytime landmark, but on the elusive objects of the night sky.

Now comes the work of setting up proper. For this, you will need to move your scope to the site—or a site—where you expect to use it: a place with a reasonably unobstructed view in all, or at least most directions—a clear view to the south is especially important because of the many interesting objects in Scorpius and Sagittarius you will want to look at—and with the ground fairly level. If you have available, as I have, an open wooden deck, you're lucky; if not, pick the most level spot you can find that is otherwise suitable.

At this point you will need a small carpenter's or handyman's level—borrow one if you must, buy a cheap one if you can. Use it to level the scope supports. With a pedestal support, this is relatively easy, since you simply check whether the pedestal pipe is vertical. On a deck—assuming the builder did a competent job and it hasn't settled since—it probably will be; on the ground, you will need some strips of wood—thin scraps of plywood will do fine—to wedge under the pedestal supports. Check with the level *opposite* each of the three supports, using your scrap wood to adjust each of them in turn, until your pedestal tube is vertical, or very nearly so, whatever side you check it on (it is perfectly possible for the tube to be vertical in one direction but out-of-true in another).

With a tripod support, the job is a little harder, since none of the three legs will be vertical. If your scope has a latitude scale on its mounting (all should, but some don't), set the scale at 0, and use your level on the tube, which should be perfectly horizontal. Turn the tube so it lines up successively with all three legs, which in this case you can adjust by varying their length, since the tripod will be (or should be) "telescopic" in both senses. If you don't have a latitude scale, you'll have to level as best you can by eye—and hope your eye is a reasonably straight one.

Once you get the scope properly leveled, it's a good idea to make certain you can get it back into position easily should you have to move it. On a deck, mark the positions of the three feet (or legs) with a pencil or ball-point; on the ground, do the same with three rocks or wooden pegs.

Now comes the really tricky part. Ideally, your scope will have a latitude scale on its altitude pivot, and a "setting circle"—a graduated dial—on the equatorial axis *which is fixed in position by the manufacturer.* If this is the case (as it is on my 2.4-inch scope), there are only two things you need do. First, during daylight loosen the altitude screw and move the tube until the scale reads the latitude of the place you're observing from, then tighten the clamp or screw as much as you can. (If you don't know your latitude, you can find it on most road maps of the area.) This ensures that your telescope will in fact point at Polaris (*i.e.,* the celestial North Pole) when turned north, since as noted on page 23 the pole's elevation varies with latitude. Now rotate the scope on its equatorial axis until the declination scale reads 90°, and clamp it.

The final step must wait until nightfall. Loosen the azimuth clamp (or screw) and rotate the scope on its azimuth pivot until you have Polaris centered in the finder scope, or as close to the center as you can get. (If it is more than a degree or so off center, better recheck your leveling and your latitude adjustment.) At this point, you should, if you've done everything right, be able to see Polaris also in the main scope, at least at low magnification.

To sum up, your scope will (or should) now, at dec 90°, be pointing toward true north, and elevated at an angle equal to your latitude—*i.e.,* your polar axis will be pointing at the celestial pole (or, what comes to almost the same thing, at Polaris).

If your elevation pivot does not have a latitude scale, *or* if your declination setting circle is not fixed in place by the manufacturer, setting up is a little more complicated. In the first case, rotate the tube so that your (fixed) declination scale reads 90°, then, using both your azimuth and altitude (*i.e.,* latitude) pivots, sight the finder and main scope in on Polaris as just described, and tighten both of them as firmly as you can. In the second case, set your latitude scale at the correct lati-

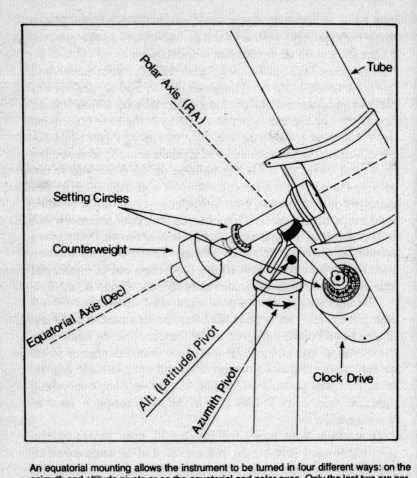

An equatorial mounting allows the instrument to be turned in four different ways: on the azimuth and altitude pivots or on the equatorial and polar axes. Only the last two are normally used, however; the first two are employed only when the instrument is first set up, then locked in place indefinitely.

tude and tighten it; then, moving the scope on the equatorial axis and the azimuth pivot, sight in on Polaris. Clamp the azimuth pivot tightly, then rotate the declination setting circle until it reads 90°, and tighten *it*.

(Incidentally, it is a good idea always to return the scope to dec 90° when you are done with your night's observations, since this will enable you to check your setup—by sighting on Polaris—next time you use it.)

If you find these instructions a little confusing, think of the problem this way: What you are after is a setup in which (1) your tube is exactly parallel to the polar axis (*i.e.*, at dec 90°), and in which (2) the altitude and (3) azimuth adjustments are such that the tube—and therefore the polar axis—then points to the celestial pole (*i.e.*, for practical purposes at Polaris). If the altitude setting (2) is known (*i.e.*, if you have a latitude scale) *and* dec 90° (1) is also known (*i.e.*, if you have a fixed dec setting circle), all you need to adjust is the azimuth (3). If only one of these is known, you can still use the "known quantity"—either declination (1) or altitude (2)—to determine the two unknowns—either altitude (2) and azimuth (3), or declination (1) and azimuth (3)—by sighting on Polaris.

If none of these three quantities is known, you have got problems; the best you can do is approximate. Begin by clamping your polar axis, then move the tube on the equatorial axis until it is as nearly parallel to the polar axis as you can estimate by eye. Clamp the equatorial axis, then rotate your declination setting circle until the pointer reads 90°, and clamp it. Now proceed as in the previous case, sighting on Polaris and adjusting altitude and azimuth until you have the star in your cross hairs (and also, of course, in the field of the main scope); clamp both altitude and azimuth. Now, as a check, rotate the tube around the polar axis; Polaris will move slightly in the finder scope, but hopefully not more than a degree or two (by this time, you should have checked the field width of your finder scope against, say, the Pointers, as you earlier did the field of your binoculars). If Polaris begins drifting too far out of the cross hairs, readjust the tube, this time using the equatorial axis and *either* the altitude or azimuth—whichever seems to work best. When Polaris is again centered, readjust your declination circle so that it again reads 90°, return the tube to its original position and recheck Polaris. It should still be on, or close to, the cross hairs; if not, you may have to repeat the process of adjustment until it remains within one or two degrees of the cross hairs *whatever the position of* the tube, when rotated around the polar axis—*i.e.*, with the equatorial axis set at dec 90°.

As a final check, sight your finder on Arcturus and then on Deneb, to see whether their declinations, as shown on the declination circle, are the same as their actual declinations (19½° and 45°, respectively). If they are more than 2 degrees off, the chances are that your mounting was improperly leveled; check it as explained on page 94 above. If the error is no more than 2 degrees, make a mental note for future reference, then clamp your declination circle and the altitude and azimuth pivots as tightly as you can; you are now set up and ready to begin using your scope.

Finding Your Targets

When Mark Twain was just beginning his apprenticeship as a steamboat pilot on the Mississippi, he complained bitterly to his boss that it was apparently necessary to learn the shapes of the river "as well as I know the shape of the front hall at home."

"On my honor," the pilot replied, "you've got to know them *better* than any man ever knew the shapes of the halls in his own house."

Unlike the Mississippi, whose sandbars, snags and channels—as the fledgling pilot discovered—shifted their positions from week to week, the heavens do not change; the positions and patterns of the stars, clusters and other objects will be the same next month, or next year, as they are today. Nonetheless, the principle holds: To use a telescope with any skill or pleasure, you need to know the stars very nearly as well as you know the objects in your front hall. For this reason, among others, it is a good idea, before examining any part of the heavens with your scope, to review its patterns, both with the naked eye and with binoculars; time spent in this manner will be saved twice and three times over when you shift over to telescopic viewing.

As a warmup, begin with some of the brighter stars, or any

planets that may be in the sky; this will accustom you to targeting in quickly on different parts of the heavens. To do this, loosen both your dec and polar (right ascension or RA) clamps enough so that the tube moves freely, move the tube until your target is in the finder scope's cross hairs, and clamp the dec axis. Now, using your lowest-powered eyepiece, bring the object into focus. What you are after is an image that is not in the least hazy (assuming, of course, that the atmosphere is not hazy) but—in the case of stars—is pinpoint sharp. With a scope larger than a 3-inch aperture, you may find it easier to focus, *not* on the brightest stars, whose images may be brilliant enough to dazzle you slightly, but on nearby, dimmer stars, at least a few of which will almost invariably be present in the same field. When you reduce *these* to pinpoints, you can be sure that your focus is as "tight" as you can get it.

Note how the image gradually moves across—and eventually out of—the field; you control this by means of your polar-axis slow-motion knob, or by cutting in your clock drive if you have one. Now try successively higher-powered eyepieces, noting that the apparent motion—unless you are using a clock drive—becomes more rapid the higher the power. Unless you happen to have a "zoom" eyepiece (and sometimes even if you do), you will have to refocus each time you change the magnification. Note that the *greater* the magnification, the *shorter* the focal length—meaning that the eyepiece lens must be moved closer to the main tube. Note also that the higher the magnification, the more delicate the job of focusing. At 50 or even 100 power, your object will come into focus more or less gradually as you move the focusing knob; at higher powers, it will come into focus so rapidly that you may easily overshoot and have to move back; learn to turn your focusing knob as slowly as possible.

Practice targeting on conspicuous objects in different parts of the heavens. You will probably find that for certain objects—those near the zenith—you must become something of a contortionist to use the finder scope. This problem is particularly serious with refractors, in which—for optical reasons—both the eyepiece and finder are mounted at the bottom of the tube, meaning that for high-elevation, near-zenith objects you will need to go on your knees. This, incidentally, is another reason for preferring a reflector to a refractor. If the latter is all you have, you may well conclude, as I did some years ago, that near-zenith objects are better examined later, when they have moved to more accessible positions.

By this time, you will be growing increasingly aware of an important difference between a telescope and binoculars: In the former, the image is somehow turned, so that even in the finder it looks quite different from the binocular image—which is simply an enlargement of what your naked eye sees. Most books on astronomy will tell you that this is because a telescope inverts the image—which is true, but only a half truth. In fact, a telescope not only inverts the image but "flips" it; in effect, it rotates the image by 180°, so that top becomes bottom and right becomes left. Attachments are available which will restore a telescope image to its normal orientation, but they should be avoided. Every unnecessary lens you

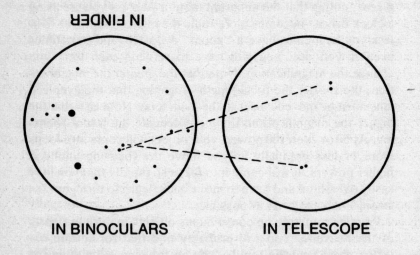

Stars seen through a finder scope resemble the same stars seen through binoculars, but with the picture turned 180°. A telescope produces the same shift in orientation, but also shows greater detail—and a much narrower field.

interpose between the objective and your eye will—owing to reflections from its surfaces—slightly reduce the light that gets through to you, which will of course defeat one of the telescope's main purposes: to gather as much light as possible.

To use a finder scope under these "inverted" conditions requires considerable practice, both in recognizing star patterns that are, as it were, standing on their heads, and also in moving the scope to bring a particular object into the cross hairs. A good tip to remember is that you need to move the scope *in the direction you want the image to move*. With binoculars—try it if you don't believe me—the opposite is the case: If a star lies at the top of your field, you "move" it *down* toward the center of the field by moving the binoculars *up;* similarly, an image too far to the left is moved *right* by shifting the binoculars *left*. Since this is the same way in which we use our eyes and heads to bring an object into view, the chances are that you never even bothered to think about it in using binoculars. With a telescope, however, you will need to form a new set of habits. When you have an object at the edge of your finder, visualize which way you want it to move so as to get it in the cross hairs and push the scope gently in the same direction; surprisingly, you will find the object apparently moving the same way as the scope.

It is worthwhile looking at some familiar, tight groups of stars through *both* binoculars and finder, in order to accustom yourself to the differences (and similarities) in what you see. For example, consider Antares. As you should be aware by this time, it has a sort of "companion"—Delta-Sco—about 2° *to the right and* (in late June, July and early August) *a little above it*, while between the two, but somewhat *lower*, lies the globular cluster M-4, easily visible in binoculars on any really clear night. In your finder, Antares and Delta-Sco will still be separated by 2°, but the latter will lie *to the left of, and slightly below*, its first-magnitude partner, while M-4 will still be between them, but slightly *higher*. Other useful groups for helping you make this mental adjustment are the tiny triangle in Lyra—Vega and its two companions, Epsilon-Lyr and Zeta-

Lyr—and the Lozenge in Draco, both of which should fit neatly into the field of both binoculars and finder. The more easily you can recognize a star grouping that has been turned 180° from its normal position, the more easily—and certainly—you will be able to find objects with your telescope.

This "trial run" period is also a good time to accustom yourself to the brightness of stars as seen through the telescope. They will, of course, look brighter than in binoculars, and much brighter than they appear to the naked eye, since the increased brightness is the main reason for using a scope in the first place. Vega, for example, will be brilliant even in a 3-inch scope, while in a 6-inch it will blaze forth like an arc lamp. Another way of looking at the matter is to bear in mind that if a star, seen through the telescope, looks like what you have become used to thinking of as second or third magnitude, it is in fact fourth to sixth magnitude—in the latter case, invisible or nearly so to the naked eye. Your ordinary second-magnitude stars, such as those in the Dipper, will appear through the scope at least as bright as Deneb to the naked eye, and in larger scopes as bright as (or brighter than) naked-eye Vega or Arcturus. Given the strange appearance of the skies through your finder scope, the more clues you can put together—of pattern, of brightness, of color—the more certain you can be of getting on target for the star you want to look at.

As you begin to be familiar with the instrument and learn to target in on the "inverted" images in your finder, you can begin looking at some of the interesting objects you already know from your binocular stargazing. Many of these are listed, and discussed in some detail, in the next chapter; here I give only a few that will do for "educational" purposes.

Apart from the moon and the planets—whose observation, through both binoculars and a telescope, is discussed in later chapters—the objects you will find most worth examining are the open and globular clusters with which you are already familiar, the diffuse nebulae that you have probably been looking at but have not seen very clearly (see page 68), and another class of objects you have seen almost nothing of with binoculars—double stars.

A good way of beginning your observation of doubles is

with the old familiar test object, the "horse and rider" in the Dipper's handle. These two stars, Mizar and Alcor, are not a true double (they could, I suppose, be classed as a widely separated visual double), but Mizar itself, the brighter of the two, shows up clearly as a true double or binary even at low magnifications, its two components being respectively of magnitude 2.4 and 4.0. Each of these stars is in turn a double, but no telescope on earth is powerful enough to show them as such; only the peculiarities of their light, as analyzed in the spectroscope, reveals their duplicity. For this reason, they, like similar stars, are called "spectroscopic binaries."

Polaris, too, is a double, but harder to see as such. Its components, like those of Mizar, are well separated, but the dimmer of them is only ninth magnitude, meaning that a 3-inch scope will show it only on a really clear night, though a 6-inch will reveal it with no trouble. Even trickier is Antares, which possesses a seventh-magnitude companion so close to it that only a 6-inch (or larger) instrument, at high magnification, has much chance of separating it from its far more brilliant companion. If you can spot it, however, note the color contrast— the pale bluish of the companion versus the orange glare of Antares itself.

For a really tough job of separation, try the "double double" in Lyra. The "double" will, of course, be familiar to you with binoculars (see page 64), but to split each of these into its components will require high magnification, and at least a 3-inch scope—in fact, if you can do it with a 3-inch, you can congratulate yourself on excellent eyesight (and excellent technique and seeing conditions!).

Not as a test of anything, but for sheer enjoyment, look at Albireo—the star marking the foot of the Northern Cross (or the beak of Cygnus—see page 29). To the eye a not especially distinguished third-magnitude star, it is revealed by even a small telescope as a double whose components are a topaz yellow and an intense, pale, sapphire blue. I am not sure whether Albireo is the most beautiful object visible in a small telescope, but I would be prepared to argue the point—as you will be, once you have seen it.

Many open clusters are among the "easy" telescopic objects. Even a 3-inch will reveal such clusters as M-6 and M-7 in Scorpius (see page 67), which in binoculars are hardly more than "grainy" patches in the field, as true aggregations of stars. For viewing in late summer, try the double cluster in Perseus (page 74), which, being brighter than either of the two just mentioned, will show even more detail. This object, indeed, is so large in the field that only at very low magnifications will you be able to see both clusters simultaneously. Note, however, that both of them have the same declination, so that you can "glance" from one to the other merely by adjusting your RA setting with the manual slow-motion controls.

The telescopic view is even more exciting when it comes to globular clusters. You will recall, for example, that M-13, in the Keystone (page 61), looks through binoculars like nothing more than a tiny ball of cotton. A 3-inch scope will show it as merely a larger and brighter ball of cotton, but in a 6-inch, if the sky is clear and dark, the "cotton" is seen to be surrounded by a delicate lacework of tiny stars, as if that part of the field had been dusted with the most finely ground diamond powder. These are the stars of the cluster's outermost layers; deeper in its structure, the component stars are set too close for resolution by a 6-inch scope—or, toward its very center, even by the largest scopes. A 6-inch scope, and in some cases a smaller one, will reveal similar structural details in such prominent "globulars" as M-4 in Scorpius (page 67) and M-22 in Sagittarius (page 69).

For viewing the diffuse light of nebulae, you will want to use your lowest magnification, at least to begin with. Of the three prominent ones in the summer sky (page 69), the Trifid, even in a 6-inch, will still look more "clustery" than "nebulous," but in the Lagoon the luminous gas clouds show up clearly in a 6-inch while in the Horseshoe they are clear even in a 3-inch. When it comes to nebulae, however, the real prize is the late summer, before dawn Great Nebula in Orion (page 79), whose clouds are brilliant in a 6-inch, while even a 3-inch will show the Trapezium, the tight cluster of four stars whose light gives the nebula its intense glow.

After the wonders your scope has shown you when examining double stars, clusters and nebulae, your first telescopic view of M-31 in Andromeda (page 71) will almost certainly come as an anticlimax. A hazy, elliptical glow in binoculars, it remains merely a larger and somewhat brighter glow in even a 6-inch scope, though a really dark, clear night may reveal traces of its dimmer outer regions. Don't expect, however, to see even a suggestion of its spiral structure in anything less than a 10-inch instrument, while resolving individual stars in it is a job for the very largest observatory scopes. It is, after all, a million and a half light-years away!

At low power, you may be able to see two other galaxies in the same field with M-31. M-32 lies only half a degree toward the south (*i.e.*, in the opposite direction from Polaris), and is in fact thought to be a satellite galaxy of M-31, comparable to the Magellanic clouds vis-à-vis our own galaxy; with good seeing, it should show up even in a 3-inch as a tiny, hazy oval. The galaxy NGC205 (it has no Messier number) lies less than a degree northwest of M-31; though larger than M-32 it is harder to see, since its structure is more diffuse, lacking the dense central nucleus that characterizes M-32; your chances of seeing it in anything less than a 6-inch scope are small.

Thus far, we have limited ourselves chiefly to objects that are "visible"—at least under good seeing conditions—in binoculars and in the finder scope, and are, moreover, distinctive enough in appearance so that you can pick them out with some certainty from the surrounding stars. What about the less visible, less distinctive objects, which either do not show up at all in a spotting scope or, if they do, cannot be picked out from other objects in the same field?

First, let us consider the borderline cases, which can be distinguished in binoculars but not, or not certainly, in your finder, whose aperture will probably be somewhat less than that of your binoculars (most finders are 30 mm or less, whereas most binoculars are 35 mm or more). One such is the globular cluster M-4, which except on the haziest nights shows up clearly in binoculars—assuming, of course, that Antares is well up in the sky—but which except on the clearest nights will probably not be spotted easily in the average finder. For this object, and for a few others of similar nature, you can try a method which usually works for me, though not always; I call it "pattern finding."

Pattern finding depends on the existence near the object in question of at least two clearly distinguishable stars; for M-4 they would be Antares and its near neighbor, Delta-Sco. Through the binoculars, these two stars, with M-4, form an almost perfect isosceles (two-sides-equal) triangle—a rather flattened one—with M-4 marking the obtuse angle where the two equal sides meet. Examine this pattern carefully through your binoculars, noting the relationship of M-4 to the two stars on either side, and its distance from the imaginary line connecting them as compared with the length of that line. Now look through your finder at the two stars, and adjust your scope so that the cross hairs form the same pattern (turned 180°, of course) with them; if your eye is good, the cross hairs should be centered on or very near M-4 even though you can't see it. Now look through the main scope at low magnification and you should be able to see the cluster; if not, try scanning back and forth, first with your RA slow-motion control, then with your dec slow motion (if you have it), and you should—if you haven't jarred the tube in the meantime—be able to pick it up.

There are, however, only a few borderline cases of this sort; for the really "invisible" objects you will need to use a different method, and one that will require some preliminary explanation. Before getting into this, however, let me say that for the beginner, at least, few of these objects will be worth bothering with at all with a small (3-inch or less) scope; though you can, with practice, find them, they won't look like much when you do.

Assuming you have a medium-sized (or larger) scope, here's how to go about it.

Finding an invisible object in the sky is a good deal like long-distance navigation at sea, where you start from a place or marker (such as a lighthouse) whose position (latitude and longitude) is known, and want to get to some other place whose position is also known. For "navigating" about the heavens, your lighthouse is a "marker" star, the more conspicuous the better; to find your "destination" object you need to know of course both its position and the marker's.

As mentioned in an earlier chapter, in the sky latitude is called declination (dec) and longitude, right ascension (RA). Declination is based on "natural" coordinates, with the celestial North Pole at dec 90°, the celestial equator at dec 0° and the celestial South Pole (which cannot, of course, be seen in the northern hemisphere) at dec −90°. Right ascension has no such natural reference points as the equator and the poles; it requires an arbitrary "zero point" comparable to Greenwich, England, on the earth's surface, whose longitude was long ago picked (by the British—who else?) as the arbitrary "prime meridian" of longitude. In the heavens, the prime meridian is called the Equinoctial Colure, (RA 24 hrs 00 min), which as it happens is defined almost exactly by the line passing through Polaris, Caph (in Cassiopeia) and Alpheratz (in the Great Square). RA increases as you move east from this line, reaching a maximum, after passing all the way around, of 23 hrs 59 min.

Suppose, now, you want to find the galaxy M-81, in the constellation Ursa Major, whose coordinates are RA 11 hrs 07 min, dec 69° 56′. The nearest useful marker star is Dubhe (the Pointer nearest Polaris), whose coordinates are RA 9 hrs 52 min, dec 62° 01′. To find M-81, you proceed as follows, step by step:

1. Using low magnification, get Dubhe precisely in the center of your field.

2. Note the declination reading for Dubhe; it should be 62°, but unless your initial setup has been unbelievably accurate, will probably be off by a degree or so. Note the amount of error, and whether it is positive (*i.e.*, the reading is greater than 62°) or negative (*i.e.*, less than 62°).

3. Recheck the position of Dubhe in your field, and if necessary adjust the scope so that it is again precisely in the center.

4. Loosen your RA setting circle, set it to Dubhe's RA, 9

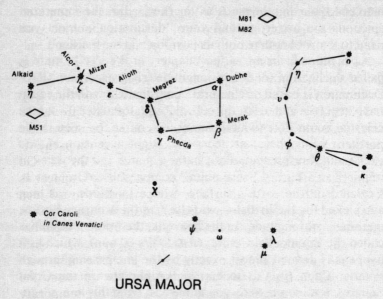

URSA MAJOR

The symbols used on this map are identified in Appendix A.

hrs 52 min (actually, given the relatively small scale of such circles, to as close to that figure as you can estimate) and tighten it again.

5. Move your tube on its polar axis until the RA setting circle pointer shows the RA of M-81—11 hrs 07 min; clamp the polar axis.

6. Move the tube on its equatorial axis until the dec pointer reads the declination of M-81—for all practical purposes 70°—*plus (or minus)* any declination error you previously noted for Dubhe (see 2 above).

At this point, if you are using low magnification, M-81 should be a tiny, blurred ellipse in your field—though you may have to scan back and forth a little with your slow-motion controls.* As a bonus after all this rigamarole, you should be able to find M-82, another relatively bright galaxy, in the same

---

*A standard reference book, supposedly written for amateur astronomers, describes M-81 as "a beautiful spiral." I daresay it is—if you happen to have to 10- or 12-inch telescope handy!

field, appearing as a considerably narrower ellipse. It lies less than a degree due north of M-81, so that it can be centered with no change in your RA control and just a touch on your dec control. Both these objects are, of course, "island universes" like our own Milky Way galaxy or M-31 in Andromeda—but much farther away than the latter.

Another interesting "invisible" object worth practicing on is M-57, the Ring Nebula, in Lyra. Its coordinates are RA 18 hrs 52 min, dec 32° 58', and our marker is, as you may have guessed, Vega, by far the most conspicuous object in the vicinity, whose coordinates are 18 hrs 35 min and 38° 44'. Go through the six steps given above, and M-57 should appear as a misty patch of light, about the size of Jupiter, in a 3-inch scope; even a 6-inch will show little of its structure, which in large telescopes closely resembles a smoke ring. It is an object of a rather rare type, called a planetary nebula—quite different from the diffuse nebulae we are already familiar with in Sagittarius and Orion. It is thought to represent the "smoke" from a cosmic explosion—perhaps the temporary transformation of a star into a nova—during which process some of the star's matter was ejected to form an encircling cloud illuminated by the remnants of the star. The latter, indeed, is still visible, but only in an observatory, being of about fifteenth magnitude. The fact that this very dim object can illuminate a large mass of gas enough to make it visible through a small scope seems paradoxical; it is explained by the star's invisible but very intense ultraviolet radiation, which by exciting the atoms of the gas cloud causes them to emit visible light, much as fluorescent paint does under an ultraviolet bulb.

Finding these and other invisible objects with your scope will not, as I have suggested earlier, show you much of visual interest; it will, however, challenge and enhance your growing skill in handling the scope. Of course, should you eventually become sufficiently fascinated by astronomy that you go on to a scope of eight- or ten-inch aperture, and especially if you become involved in sky photography, an ability to use the scope in this fairly sophisticated way will come in very handy.

By that time, however, you will have graduated well beyond the level of this book.

Care of the Telescope

Most telescopes are tolerably rugged instruments, and are unlikely to sustain damage unless they are dropped heavily or tipped (or blown) over. A much more serious, though far more insidious, danger is moisture, which can corrode the metal working parts. It can also—and in fact almost invariably will—complicate viewing by condensing on the optical elements in the form of dew. A good stargazer's night is also, of course, a clear night, meaning one in which the earth, and objects on it, cool off relatively rapidly—more rapidly than the atmosphere. The result is that, except in the driest regions, atmospheric moisture will sooner or later condense on the lens of the eyepiece and (in refractors) the objective, blurring the view with a film of moisture.

The dew problem on a refractor objective can be minimized, though not eliminated, by constructing a "dew shield"—a one-foot tube of rolled black paper, preferably shellacked or varnished to keep it from absorbing moisture—which can be tied to the objective end of the refractor, thereby excluding much (though not all) of the air that brings moisture.

When dew accumulates—as it will, sooner or later, if the telescope remains out of doors on a clear night—it is important *not* to try to wipe it off the lenses or mirror, which may well scratch them; instead, simply suspend observations for that night, cover the scope with a plastic tarp—or even an old blanket—and leave it. (Alternatively, of course, you may prefer to take it indoors.) Next day, simply by removing the cover, you will allow the moisture to evaporate with no traces. Should you for any reason *have* to wipe any of the lenses—if, for example, some wicked child (or adult) has left fingerprints on them—use lens tissues (sold in most drugstores) or, at the very least, a *clean* handkerchief, meaning one that has not had a chance to pick up dust or grit in somebody's pocket. It is a good idea to blow on the lens gently before wiping it, to remove any dust or grit that may be present.

A little 3-in-1 Oil sprayed or wiped onto the metal parts will protect them against corrosion; this may be done once at the beginning of the season, and should in any case be done at the time you put your scope away for the winter. With even a minimum of care, a telescope should last for years. In the case of reflectors, the mirror may need to be resilvered from time to time; how often depends on atmospheric conditions and other things. At worst, however, this should not be necessary more than once in three to five years, and with luck not half that often. Consult the manufacturer (for American-made instruments) or one of the astronomy magazines for information on how to go about this.

Chapter 5

THE TELESCOPIC SKIES
Some Things to Look For

BY this time you will—or should—have a fairly detailed acquaintance with the more interesting celestial objects that can be seen through binoculars, as well as at least a rudimentary knowledge of how to find objects with the telescope. What follows is a guide to *some* of the interesting things you can see with a telescope, including both the binocular objects already discussed in Chapter 3 and more elusive ones that show up well only in a telescope. These include both open and globular clusters, a modest number of nebulae which can be seen in a small scope, and a sizable number of the more easily found double stars, almost none of which (as already noted) look like doubles in anything but a telescope.

I should emphasize that I have *not* tried to give a comprehensive amateur astronomer's guide to the heavens, listing every object, both obvious and obscure, that can be seen, however dimly, with a 3- to 6-inch scope. Rather, I have limited myself to the main "points of interest"—with only a few exceptions, only "Messier" objects* and not all of those. Working your way through even this curtailed list should occupy you for two or three summers at least, by which time you will be well prepared to graduate to more detailed guides and more difficult objects.

*See Appendix A.

Most of these objects should be findable with the aid of the accompanying diagrams and your finder scope; for the more elusive objects, I have in each case given a marker star which should enable you to locate it. You will find the coordinates of the markers listed beginning on page 180, and those of the objects themselves in the catalogue of Messier Objects beginning on page 183; with these in hand, you can proceed by the method for finding "invisible" objects described in the previous chapter.

Beginning with the circumpolar constellations, visible at any time during the summer, we find that Ursa Minor, the Little Dipper, is of no special interest, apart from Polaris itself, which as already noted is a double star with a ninth-magnitude companion. The main star of the two is also double, but being a spectroscopic binary cannot be seen as such even with the largest scopes.

Ursa Major is rather more interesting, but most of its points of interest—the double star Mizar and the two galaxies M-81 and M-82—have been mentioned already (see page 108). Another, rather dimmer galaxy, M-51, lies about 4° southeast of Alkaid, at the tip of the dipper handle, which can serve as your marker star if you have difficulty finding it. It is called the Whirlpool Nebula—but this, of course, refers to its appearance in a large telescope.

Note, by the way, that here and elsewhere in this chapter, "south" does *not*—or not necessarily—refer to the direction of the southern horizon, but to the *direction opposite to that of Polaris*. It may help if, when you look at any of your marker stars, you visualize a line connecting it with Polaris; that line will then run north and south, and one at right angles to it, east and west.

Draco, like Ursa Minor, is of little telescopic interest, but Cepheus is more attractive. Two of its main stars, Beta-Cep and Delta-Cep, lying at opposite corners of the "house," are double, and another, fourth-magnitude double, Xi-Cep lies almost exactly in the center of this figure.

Turning now to the early-summer constellations, we find that Gemini possesses a conspicuous double in Castor—the

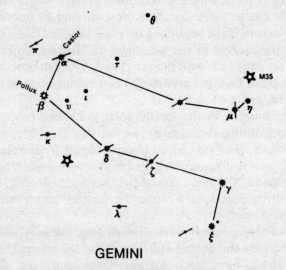

GEMINI

The symbols used on this map are identified in Appendix A.

brightest binary star in the northern hemisphere. Gemini also contains a bright and attractive open cluster, M-35, but it is not easy to track down in summer. Even in early June, it is already below the horizon at nightfall, while in late summer it does not rise until the morning hours. If both Aldebaran and the Twins are visible (*i.e.*, before dawn in late summer), M-35 can be seen in binoculars about halfway between them. At other times, it can be found either by reference to Aldebaran and Capella, with which it forms an almost equilateral triangle, or—using the "invisible" method—with the help of El Nath as a marker.

Cancer of course contains the Beehive, a striking cluster at low power. Another, less conspicuous cluster, M-67, lies about 8° almost due south of the Beehive. To find it with binoculars, note the two fourth-magnitude stars in the same field with the Beehive and east of it, and follow their line south. Alternatively, use Regulus as a marker.

Leo contains a conspicuous double star, Algieba, which,

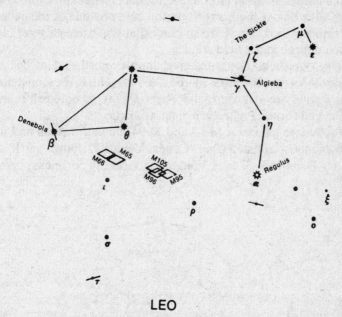

LEO

The symbols used on this map are identified in Appendix A.

like Regulus, forms part of both the Trapezoid and the Sickle; at second magnitude, it is the brightest star in the Sickle apart from Regulus itself. Denebola appears in the telescope as an attractive blue-orange pair, though it is only a visual double. In the region between Denebola and Regulus lie no less than five galaxies, M-65, M-66, M-95, M-96 and M-105; all are rather closer to Denebola, which is therefore the best marker for them.

East of Leo, in Coma Berenices and the adjacent regions of Virgo to the south, are far more galaxies—in fact a galactic cluster*; the cluster includes M-60, M-94, M-99, M-100 and M-84 through M-88, plus the two galaxies mentioned in the next paragraph but one. Coma Berenices also includes the globular cluster M-53, of about eighth magnitude. For all

*Comparable to a star cluster, but much, much bigger.

these objects, Denebola is the most convenient marker, with Spica almost as good later in the summer, when the Lion's tail has fallen below the horizon. In the case of some of the galaxies, moreover, several are so close that you can easily get two or even three in the field at once.

Virgo contains two attractive doubles northeast of Spica: Gamma-Vir, at the fork of the "Y" in which the constellation's stars are arranged, and Theta-Vir, lying between it and Spica and forming with them a flat triangle.

Two more galaxies, M-63 and M-94, lie in the small and inconspicuous constellation Canes Venatici, lying north of Coma Berenices. Canes Venatici is virtually formless, being

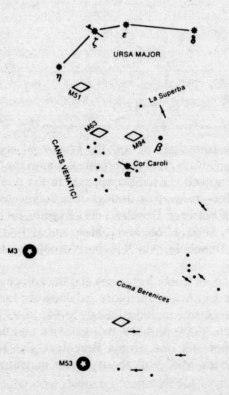

The symbols used on this map are identified in Appendix A.

composed almost entirely of fifth- and sixth-magnitude stars, but can be located easily through its main star, third-magnitude Cor Caroli. Taken with the Spring Triangle of Arcturus, Denebola and Spica, Cor Caroli forms an almost perfect diamond (sometimes known, in fact, as the Diamond of Virgo); it lies in about the same position *north* of the line Arcturus-Denebola as Spica does to the south, and is also a well-known and easily resolved double. M-63 and M-94 lie respectively northwest and northeast of Cor Caroli, and within about 4° of it; Alkaid in the Dipper is the best marker for them. Canes Venatici also contains the bright globular cluster M-3, lying on the line from Cor Caroli to Arcturus but nearer the former. A less conspicuous but more unusual object is the star La Superba, lying about 7° north and a little west of Cor Caroli. Only fifth magnitude, it is known for its brilliant red color—one of the few truly red stars in the heavens—which, however, requires a telescope to be visible in its full beauty.

Boötes contains two fine doubles, Epsilon-Boo and Delta-Boo, of third and fourth magnitude respectively; they lie northeast of Arcturus and form almost a straight line with it. The first is a tight (separation only 3″) orange-green pair, the second, a distant (separation 105″) yellow-blue pair.

Corona Borealis has no attractions for the telescope; you will have to be satisfied with its naked-eye beauty—which is not to be sneezed at.

Hercules, of course, contains the famous globular cluster M-13 which we have already mentioned (see page 61), and also the less conspicuous globular, M-92, lying above the Keystone; Eltanin, in Draco, is the best marker for it. On the opposite side of the Keystone lie two third-magnitude doubles—Delta-Her, a greenish-purplish visual double, and Alpha-Her, an orange-greenish pair.

Half a dozen globular clusters lie south of Hercules, in the companion constellations Ophiuchus and Serpens. They are not easy constellations to describe, but the principal (and marker) star for the area, Rasalhague, can be found easily with the help of the Summer Triangle, with which it forms a kite-shaped figure, occupying about the same position south-

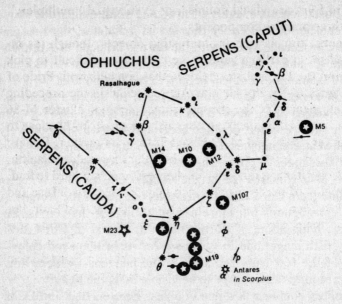

OPHIUCHUS

SERPENS (CAPUT)

SERPENS (CAUDA)

Rasalhague

M14 M10 M12

M5

M107

M23

M19

Antares
in Scorpius

The symbols used on this map are identified in Appendix A.

west of the line Vega-Altair as Deneb does to the northeast. Using this star as a marker, you should be able to locate the globulars M-10, M-12 and M-14. M-5, another globular, is far enough west so that Arcturus will serve as well or better, though at brighter than fourth magnitude, binoculars should locate it if you know where to look—just about halfway from Rasalhague to Spica. Two more globulars, M-107 and M-19, are far enough south to make Antares, rather than Rasalhague, the best marker.

For its size, Lyra is rich in interesting objects, but the two most interesting have already been mentioned: the "double double" near Vega, and the Ring Nebula, M-57. Another double is Zeta-Lyr, which marks the junction of the triangle and parallelogram that make up the main pattern of the constellation. Its components have been described as topaz and green, which I take to mean yellowish and greenish. This star, and the double double, are true doubles or binaries; several other stars in Lyra are visual doubles, or even visual "multiples," with three or more components.

Cygnus, too, is rich in interesting objects, though (as indicated in an earlier chapter) they are often difficult to pick out from the brilliant star fields in that constellation. Pride of place goes to the double star Albireo, cited in the preceding chapter; near it is the inconspicuous globular cluster M-56 (Vega is its best marker). Open clusters include M-29, close to its marker, Sadr, and M-39, about 7° from *its* marker, Deneb; another, M-71, lies in the neighboring constellation Sagitta, with Altair its best marker—its dimness makes it hard to find. This region of the sky is also rich in nebulae, both diffuse and planetary, especially in the areas around Deneb and Sadr, but few of them can be singled out from the surrounding star fields with a small scope. The largest of them, M-27, is not in Cygnus at all, but in the adjacent and obscure constellation Vulpecula (Altair is your best marker). It is so dim, however, that you will need a really dark night, a 6-inch scope and low magnification to see it at all, while your chances of seeing the nebula's structure, which gives it its name (the Dumbbell) are even dimmer. Cygnus completes the list of early-summer (more precisely, early evening in early summer) constellations worthy of telescopic note; we can now proceed to those visible later in the season—or the evening.

Of these, pride of place must certainly go to Scorpius. Its binary stars include Antares itself, though (as noted earlier) its close, dim, bluish companion can be seen only under optimum conditions—in particular, when Antares is at its maximum elevation above the horizon (*i.e.*, it lies due south, or exactly opposite to Polaris). Beta-Sco, uppermost of the "claw" stars, is a much easier double, while high magnification, and at least a 6-inch scope, will—with luck—show its brighter component as also double, possessing a ninth-magnitude companion. Nu-Sco, another complex star system nearby, appears as a double at low power; high power—and a 6-inch scope—will split its dimmer component into two, while the brighter can also be split, but only with an 8-inch-or-better instrument. Clusters include the globulars M-4 and M-80; even a small scope will show many of the constituent stars in the first of these, while the second cannot be resolved into its components even with a 12-inch. Another, more or less "intermediate" globular, M-62, lies on the opposite side of Antares, which serves as its marker as it does for the other two. Open clusters include M-6 and M-7, near Shaula—both easy binocular objects and fine sights in a telescope.

Rising later than Scorpius, Sagittarius is even richer. As you will recall from your binocular explorations, many of its most conspicuous objects are easily spotted even with the naked eye; indeed, Sagittarius is one constellation in which several of the most useful markers are not stars but nebulae. M-20, the Trifid Nebula, will easily guide you to the open cluster M-21, only a couple of degrees away, as well as the more distant M-23 to the northeast, while M-17, the Horseshoe Nebula, will similarly lead you to the nearby open clusters M-18 and M-24, as well as the more distant M-25 to the southeast. For the globular clusters M-22 and M-28, either M-8, the Lagoon Nebula, or the star Nunki, in the middle of the easternmost "bow," will serve; the latter will also help you find the rather isolated globular, M-55. Three other globulars, M-54, M-69 and M-70, lie between the lower ends of the two bows; either Nunki or Kaus Australis, at the lower end of the western bow, will serve as a marker—assuming, of course, that the southern horizon is good and clear, or that you are observing from the southern tier of states; Sagittarius, like Scorpius, conceals many of its riches from northerly observers. Incidentally, if you should need reminding, don't leave Sagittarius without taking a good look at the three marker nebulae themselves—especially the Horseshoe, which shows up by far the most clearly in a small telescope.

Moving from the southern to the northern skies, Cassiopeia contains a sizable quota of interesting objects, including two conspicuous doubles. Schedar, marking the bottom of the right-hand "V" in the constellation's "W," a widely separated reddish-bluish pair; the blue star in this visual double is only ninth magnitude. About 2° northwest of Schedar is

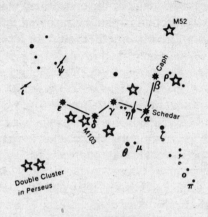

CASSIOPEIA

The symbols used on this map are identified in Appendix A.

Eta-Cas, a binary whose components have a yellowish and purplish cast. The open cluster M-52 lies almost in line with Schedar and Caph, about 7° northeast of the latter; another, M-103, lies about 1° west, and a little north, of Delta-Cas, at the bottom of the eastern "V." An even brighter open cluster—oddly enough, with no Messier number—lies in the opposite direction, about 2° from the star.

In Pegasus, only two objects are of much interest. One is the star Enif (Epsilon-Peg), a visual "multiple" with four widely separated components. It lies almost in line with Algenib and Markab, which mark the southern side of the Great Square, but well to the east of both; it is, as it happens, the only star in the vicinity brighter than any of the stars in the Square except Alpheratz. Using Enif as a marker, you should have no difficulty in finding, even with binoculars, the bright globular cluster M-15, about 4° to the northeast. Enif will also serve as a marker for the bright globular M-2, lying to the southeast, in Aquarius.

Andromeda, lying east of Pegasus and therefore rising a little later, is distinguished chiefly for the two galaxies, M-31 and M-32, described on page 105—the former, many people

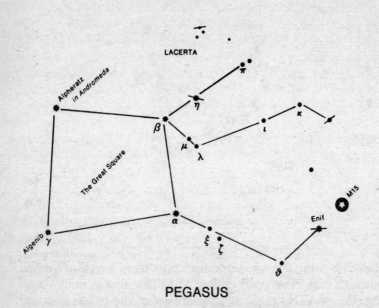

PEGASUS

The symbols used on this map are identified in Appendix A.

would say, being distinction enough for any constellation. Unfortunately it takes a considerably bigger telescope than any you are likely to have to show detail even in M-31—though some people claim to be able to see at least a hint of the star clouds beyond its bright central nucleus in a 6-inch at low power. Try it some really dark night, when your eyes have reached maximum dark adaptation, and see. Almach, lying at the end of the line of bright stars stretching northeast from Alpheratz, is a well-known bright double; its components are golden and greenish-blue.

Perseus, east of Andromeda, is of course mainly distinguished by its famous Double Cluster, which is even more striking in a telescope—at low power—than in binoculars. Another fairly bright open cluster, M-52, lies about halfway between Algol, the "demon star," and Almach, the double cited in the preceding paragraph. Several of the less conspicu-

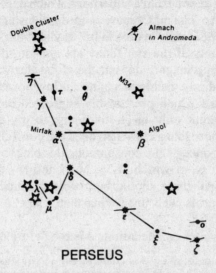

PERSEUS

The symbols used on this map are identified in Appendix A.

ous stars in Perseus, of third and fourth magnitude, are easy doubles.

Consider now the late summer and early fall constellations. Taurus, with its two close and conspicuous clusters, the Pleiades and Hyades, is outstanding—but both these objects are much more striking in binoculars than through the telescope. A low-power view, however, will at any rate suggest the sizable number of stars in both groupings which show up dimly or not at all in binoculars; you will have to scan back and forth with slow motion to cover the full area of either cluster. Aldebaran is a visual double; its companion is only tenth magnitude, but is well enough separated to make it clearly visible in a 6-inch or even smaller. Well to the east, lying between El Nath (marking the tip of the bull's northern horn) and Zeta-Tau (marking the other horn tip), but much closer to the latter, is one of the most famous celestial objects, M-1, the Crab Nebula. Though usually classed as a diffuse nebula, it is perhaps more accurately termed a super-planetary nebula. This giant cloud of gas is thought to be the expanding debris from a supernova which exploded A.D. 1053; according to ancient Chinese records, the star became so brilliant as to be clearly visible in full daylight for several weeks!* The remnants of the star today are almost invisible, being of only fifteenth magnitude, but the cloud thrown off by the explosion is dimly visible in a 6-inch at low power. The filamented structure which presumably suggested its name is, however, apparent only in photographs; to my eye, at least, it looks no more like a crab than does the constellation Cancer.

Auriga contains no conspicuous doubles, but a number of open clusters, of which M-36, M-37 and M-38 are most conspicuous. All three should be visible in binoculars when the constellation is well up; failing that, Capella can serve as a marker.

Of the winter constellations, visible before dawn in late Au-

---

*To be more precise, the explosion was *seen on earth* in 1053; it actually occurred some 3,900 years earlier, about the time the Egyptians were starting work on the pyramids, with its light taking that long to reach our solar system.

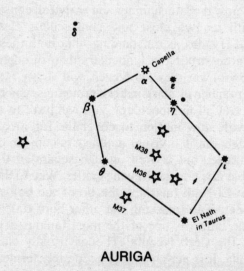

AURIGA

The symbols used on this map are identified in Appendix A.

gust and September, the champion *hors concours* is Orion. Indeed, the mighty Hunter is unique among the constellations in being equally spectacular whether viewed with the naked eye, with binoculars (page 78) or, in detail, with the telescope. The first sight of the Great Nebula, with the Trapezium blazing at its heart, through a 6-inch is something I truly envy every reader; if any celestial object is breathtaking, it is this one. Less than half a degree from it, and with the same RA (just shift your dec control slightly), is a smaller diffuse nebula, M-43, but you will be doing well if you can distinguish it from its far more brilliant companion. Another nebula, with no Messier number, lies within a degree or so of Alnitak, easternmost of the "belt" stars and yet another, M-78, lies farther to the north. Just south of Alnitak lies an almost unique object, the so-called Horsehead Nebula—a cloud of dark dust, which, silhouetted by bright dust and gas clouds around it,

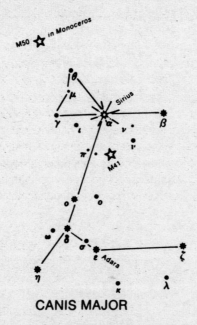

CANIS MAJOR

The symbols used on this map are identified in Appendix A.

should appear as a tiny but almost perfect horse's head; look for it in a 6-inch, at low power. For all these objects, Alnitak will serve as a marker; it is, incidentally, a tight binary while Mintaka, at the other end of the belt, is an easily resolved visual double.

Just before dawn, you may catch a sight of Sirius, spectacular even in a small telescope. About 4° due south of it is the brilliant open cluster M-41, visible in binoculars and even, under favorable conditions, with the naked eye. Realistically, however, unless you are observing in October or later, you are unlikely to get much of a look at it, since by the time it is well above the horizon dawn will have overtaken it—and you. Almost 10° northeast of Sirius lies a dimmer open cluster, M-50, in the constellation Monoceros.

Chapter 6

THE MOON IN JUNE
Or July, or August

BEFORE getting into the substance of this chapter, I think I should say quite frankly that the moon is not my favorite celestial object. Part of the reason, I suspect, is my feeling that this satellite of ours was terribly oversold to the American public during the so-called Race to the Moon of some years back.* What Neil Armstrong—well briefed beforehand, I am sure, by NASA's public relations experts—called "a giant step for mankind" was, so far as I was concerned, a giant step primarily for the U.S. aerospace industry, which profited to the tune of several billions from this exercise in cosmological one-upmanship. For the rest of mankind, the moon landing was primarily a TV spectacular, whose undeniable entertainment value was—in my eyes—considerably dimmed by the fact that it had cost me something like a thousand dollars in taxes—pretty steep admission even for the best show in town. For me, m-o-o-n still spells "rip-off."†

Quite apart from this admittedly prejudiced view, however, is the plain fact that the moon, under close examination, tends

---

*In fact, "race" is probably a misnomer, since the Soviet government was apparently not very interested in "beating" us to the moon. This suggests that even governments can sometimes act sensibly.

†One should distinguish, however, between the *manned* space shots, which soaked up about eighty percent of NASA funds, and the various robot satellite programs, which have brought in at least eighty percent of the space program's scientifically useful results—a difference, in information-per-dollar-spent, of something like sixteen to one.

to repeat itself. As everyone knows, much of its surface is covered by the craters, of various sizes, left by billions of years of meteorite bombardment.* And while it is by no means true that when you've seen one lunar crater you've seen them all, any representative sample of (say) a dozen craters will look remarkably like the next dozen. And the next, and the next. Nor does stepping up the magnification help much. Certainly it enables you to see *more* craters—but a crater 50 miles in diameter, viewed at 100× magnification, still looks very much like a 100-mile crater viewed at 50×.

Nonetheless, though the details of the lunar surface are repetitious, they are present in enormous profusion—for which reason alone they are worth looking at. The moon is, of course, far closer to us than any other celestial body—a mere quarter-million miles as against (say) 25 million for Venus at its closest, and 35 million for Mars at its most favorable oppositions.† Even before the lunar landings, we knew much, much more about the surface appearance and structure of our satellite than about any other object in the solar system—or out of it, for that matter.

It is this very concentration of detail in a very small area that produces the first major difference between moon-watching and star-watching: the critical importance of optical aid. As we already have noted, you can see thousands of stars with the naked eye, and tens of thousands—along with clusters, nebulae and a few galaxies—with binoculars. On the moon, by contrast, the unaided eye sees only a bare handful of the major surface features, while even binoculars will at best resolve no more than a few dozen; even a moderately rewarding look at the moon requires a small telescope. This is not to say that viewing the lunar surface through binoculars is pointless; at the very least it will help you fix in your mind the main "moon marks"—comparable to the sky marks discussed in Chapter 2—whose shapes and patterns will help you find your way with a telescope to less conspicuous features.

---

*A few may have been produced by volcanic eruptions.
†See Glossary.

There is one other point about the moon that ought to be mentioned: its brightness. Though the moon is far less luminous than the sun, it is far more luminous than any other celestial object; the magnitude of the full moon is about -10, which when translated by the mathematics of magnitude (see Glossary) means that it emits something like ten *thousand* times as much light as Arcturus or Vega. When "amplified" by even a 3-inch telescope, this is enough to dazzle you, while in a 6-inch instrument it will be uncomfortable and—with prolonged viewing—even damaging. The problem is most severe when the moon is at or near full, and viewed at low power, in which all or nearly all the lunar disk is pouring its light through your eyepiece. One solution is to use your solar cap—if you have one (see next chapter)—which cuts down the effective aperture of your scope; another is to buy a lunar filter—a disk of colored glass which fits into or over the eyepiece. One or the other arrangement is essential, for both comfort and safety, if you plan on giving the moon much more than a brief glance; with a 6-inch scope, it is essential anyway.

"Learning the moon" as Mark Twain learned the Mississippi River is both easier and harder than learning the stars. It is easier because there is much less to learn, harder because what you are learning changes from night to night. As you already know, the appearance of the stars, apart from its slow, four-minutes-a-day shift from east to west, is the same on Tuesday or Wednesday as it was on Monday; moreover, barring a partly cloudy sky (when you probably won't be doing much sky-watching in any case), all your sky marks are in view at once. The moon, as a thin crescent, will show only a few of its landmarks; only in the period around full moon will you be able to see all of them together.

Moreover, the very appearance of the landmarks changes, due to the changing angle of the sun relative to the moon, and the resulting change in the length of the shadows that "bring out" our satellite's features. At full moon, indeed, much detail on the lunar surface is washed out by the lack of shadow—comparable to high noon on earth—so that while full moon is a good time for getting an overall view of lunar geography—the technical term is "selenography"—the moon's individual features, with few exceptions, are seen more clearly at other times of the month. Specifically, lunar details show up most clearly near the "terminator"—the shadow line separating the satellite's illuminated and dark portions—where the lengthened shadows (comparable to those at sunrise or sunset on earth) emphasize even small differences in elevation and surface texture.

This matter of illumination can be confusing in itself, since many features that show up sharply at one time of the month will become all but invisible later on. A further confusion arises when—as frequently happens in astronomy books—several photos taken at different times of the month are joined up to produce a pictorial guide to the moon's overall features. Superficially, such composites look like the full moon, but show most surface features much more clearly than they would ever appear at that time. This, of course, is the whole idea, but it can be confusing if the novice moonwatcher compares the photo with the actual full moon as seen through binoculars or telescope—and finds he cannot see most of what the photo shows.

A final word about lunar directions. For practical purposes, directions on the moon are the same as in the rest of the heavens; north is the edge of the disk facing (or closest to) Polaris, with south opposite, east to the left and west to the right. Putting it another way, when the moon is highest in the sky, bearing due south, north is at the top and south at the bottom. These, however, refer to the moon as seen by the unaided eye or in binoculars; in a telescope, the whole business is as usual rotated 180°, so that north is on the bottom, south at the top, and east and west to the right and left respectively.*

Just after new moon, the satellite is illuminated by the sun in a narrow crescent along its western edge; at the same time,

*Another possibly helpful way of thinking about lunar directions is that when the moon is in the south, and seen through binoculars (or with the naked eye), north and south are in the same positions as on an ordinary map of earth, while east and west are reversed. Through a telescope, however, east and west are as on earth maps, while north and south are reversed. However you choose to think about them, getting used to directions on the moon will require a certain mental adjustment.

the rest of the moon is dimly lit by light reflected from earth, producing the phenomenon summed up in the poetic phrase "the old moon in the new moon's arms."

At this point, the moon will be visible—if you know where to look—during most of the day (you may have to block the sun's light with your hand, or a piece of cardboard, to see it, however) and will set soon after sunset. As the month progresses, it will set some fifty minutes later each night, while the illuminated portion of the disk gradually spreads from west to east. At first quarter it will be half illuminated and half dark, and will be well up in the southern sky at sunset; by full moon, it will be clear of the eastern horizon by sunset and will remain in the sky all night. Thereafter, darkness will engulf the moon day by day, as light previously expanded over it, beginning at the west and moving east, until a day or two before new moon the satellite, rising a little before sunrise, will show only a thin crescent on the east, facing the rising sun.

Seen at full moon—the only time that the moon's features can be seen all together—the most conspicuous surface markings are the dark areas called mares (pronounced MA-rays) or seas. They are not seas, of course—as everyone knows, the moon has no surface water (and little if any even beneath its surface)—but were so named because their darker color gives them much the look of the bodies of water shown on maps of earth. They cover most of the moon's northeast quarter and much of its northwest quarter as well, with tongues extending well into the southern hemisphere. The two groups of mares, western and eastern, are connected by two narrow arms, one lying near the northern edge, between which lies an enormous but narrow "island," stretching more than halfway across the disk.

Not only do the mares contain no water, but there is no reason to suppose they ever did—that is, they are not (as one might be tempted to think) the beds of oceans dry for millions or billions of years. (In fact, modern oceanographers are well aware that the beds of the earth's oceans, were they visible, would appear nothing like as smooth and level as do the mares.) Exactly what the mares are is uncertain. Many of them, from their shape and other indications, appear to be the remains of craters, but enormously larger than other lunar craters, and seemingly filled almost to the brim with something like lava—molten rock that has since solidified. This would imply that the mares were formed long ago, at a time when the interior of the moon was presumably still hot and liquid. As against this, however, is the fact that the mares show far fewer craters than do other parts of the lunar surface, suggesting that they were formed relatively recently, meaning that the question of how they were formed is still, like the moon itself, very much up in the air.

Less conspicuous but still prominent lunar features are long ranges of mountains formed, like the mares, by processes we can only guess at. Some of them, at least, appear to represent portions of the encircling walls of enormous craters, the remainder of which were destroyed by subsequent meteorite bombardment or other processes.

By all odds the most pervasive features of the lunar surface, however, are the craters, of which even a small telescope reveals over a thousand, their diameters ranging from 150 miles to a mile or so. Larger telescopes, and more recently lunar rockets and human explorers, have raised the number to millions, their sizes ranging down to a foot or less. Since the moon, as everyone knows, has no atmosphere, even the tiniest meteoric particles will strike its surface instead of being burned up as "shooting stars"; the scars they leave on landing, in the absence of wind or rain, persist almost indefinitely—unless, of course, they are obliterated by later collisions.

A typical crater consists of a ring of mountains surrounding a relatively smooth circular plain—which may, however, show numbers of smaller craters. The larger the crater, the higher its encircling mountains—but the smaller the crater, the higher its mountains *in proportion to* the diameter. Many craters possess isolated peaks (sometimes several peaks) in the centers of their plains, which are thought to have been formed by a sort of bounce-back from the original meteoritic impact. (Similar formations have been produced in the laboratory, by shooting pebbles or beebee shot into thick mud.) Other craters show no central mountains, suggesting that they were formed at a time when the satellite's materials were more plastic, so that the central peaks could slump back over some millions of years. Some craters seem to be filled with lava or some similar material—as may (or may not) have occurred with the mares.

A few craters show a peculiar feature called "rays"— straight, light-colored markings extending outward in all directions, for distances ranging from a few tens of miles to more than a thousand. These can hardly be anything but "splash marks"—material ejected by the original impact that formed the crater. But why the splashes are of so distinctive a color, and why so few of the moon's craters possess them, are questions no one can yet answer with any certainty. Many

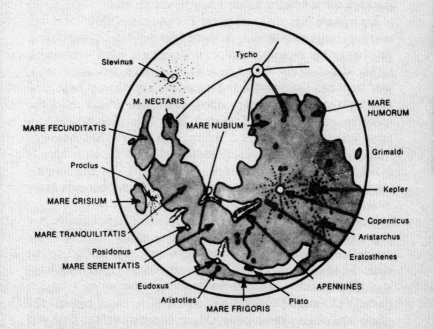

Full moon (about 14⅓ days after new moon). This and the succeeding moon maps show our satellite as seen through a telescope; for the view through binoculars, they must be turned upside down.

rayed craters are among the moon's largest; the rays—though not the craters themselves—are, like the mares but unlike most other lunar features, best visible at full moon. For this reason, and because they are both large and distinctive, the rayed craters are among the most useful moon marks in learning the satellite's overall geography.

When we examine the full moon's major features, beginning on the west, perhaps the most conspicuous near its western edge is Mare Crisium, small, isolated and almost circular; just east of it is the small rayed crater Proclus. To the south, almost straddling the lunar equator, is the larger and more irregular Mare Fecunditatis, with the large rayed crater Langrenius lying just west of it. South and west of Mare Fecunditatis lies a small, irregular and rather ill-defined "sea," Mare Nectaris, with the conspicuous rayed crater Stevinus lying between it and the moon's southwestern edge.

Both Mare Nectaris and Mare Fecunditatis "flow" through narrow straits into the larger Mare Tranquilitatis, lying on and just below the moon's equator; this, in turn, joins—through a rather wider strait—Mare Serenitatis, which lies about midway between the equator and the moon's northwest edge, and just west of the imaginary north-south line which splits the moon into an eastern and western half. Mare Serenitatis is almost circular, like Mare Crisium, but larger. Southeast of Mare Serenitatis, almost straddling the north-south line, is Mare Vaporum, the smallest—and darkest—of the moon's "seas." North of it is the narrow strait which connects Mare Serenitatis with the larger Mare Imbrium.

Moving northeast, to the region near the lunar equator, we come to two of the most conspicuous moon markings, the magnificent rayed crater Copernicus and, still farther to the east, the smaller but hardly less conspicuous rayed crater Kepler. North and east of them stretches the moon's largest "sea," Oceanus Procellarum, which terminates toward the north in two irregular gulfs, Mare Nubium on the east and Mare Humorum in the center. Almost due north of the latter is perhaps the moon's most extraordinary feature, the spectacular rayed crater Tycho. The floor of its central plain is al-

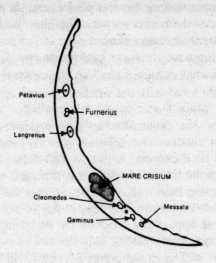

Crescent moon, almost 4 days old.

most white, surrounded by a dark circle that is its ring wall. Its brilliant rays extend almost to Stevinus to the west, almost to Kepler on the southeast, while the longest of them stretches west of north almost to the edge of the moon's disk, en route cutting across Mare Serenitatis almost like a chalkmark on a blackboard.

With these moon marks, you should not have much difficulty in finding through your telescope some of the moon's interesting but less conspicuous features. Beginning three or four days after new moon, you will see Mare Crisium as a conspicuous "blot" on the lower part* of the young crescent. Southeast and northwest of it are several sizable craters, while several smaller but sharply defined ones lie within its boundaries. South of it, on or near the terminator, lie three large craters—Langrenius, the rayed crater mentioned earlier, and above it, Petavius and Furnerius. All three are more than 50 miles in diameter (Petavius is about 100 miles across), all

*As seen through a telescope, of course.

three have conspicuous central peaks, and all three, because of the angle at which you are viewing them, will look more or less elliptical rather than circular.

In another day or so, you should be able to see Mare Fecunditatis (somewhat diamond-shaped) above Mare Crisium, and farther to the southeast the smaller Mare Nectaris. Note on the floor of Mare Fecunditatis the smallish, "twin" craters Messier (yes, the same Messier who produced the Messier Catalogue of clusters and nebulae—see Appendix A) and, to its right, W. H. Pickering, with two well-defined rays stretching east from the latter. Messier, it is thought, was formed by a meteor coming in at a sharp angle, so that the "splash" was focused in one direction. The central portion of the "splash" however, was apparently blocked by Pickering (presumably already in existence), leaving only the two rays passing along either side of it. The rayed crater Stevinus will be clearly visible (though for the best view of its rays you will have to wait until near full moon), while below it, on the southern (upper)

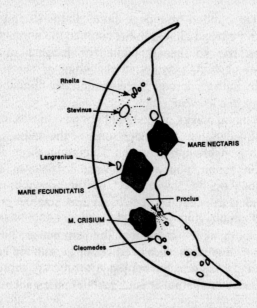

Crescent moon, about 5¾ days old.

edge of Mare Fecunditatis, is the smaller but well-defined rayed crater Orus.

Southeast of Stevinus is a curious formation called the Rheita Valley. Though its overall shape is obscured by craters that have impinged on it (notably, the large crater Rheita at its northern end), it looks like nothing so much as a gouge scooped out of the moon's surface—conceivably a sort of "bullet graze" from a high-velocity meteor that just skimmed the surface.

When the moon is five to six days old, you will be able to see, east of Mare Crisium, most of Mare Tranquilitatis and much of Mare Serenitatis, while the sun's slanting rays will reveal that the strait between them is formed by two mountain ranges, the Taurus to the west and the Haemus to the east. Between them, in the middle of the strait, lies the crater Pliny. Between Mare Tranquilitatis and Mare Nectaris to the southwest lies the large crater Theophilus, 64 miles across, with

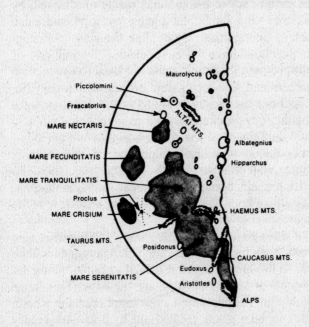

First quarter moon, about 7¾ days old.

walls up to 18,000 feet high; its high central peak can be seen within the moon's shadowed portion for almost a day before the floor of the crater is illuminated. Southeast of Mare Nectaris lies another mountain range, the Altai.

At the age of seven to eight days, with the moon at or near first quarter, Mare Serenitatis will be completely visible, along with the strait separating it from Mare Imbrium to the east. This, like the strait at the other end of Mare Serenitatis, is composed of two mountain ranges, the Caucasus to the north and the Apennines to the south. North of the Caucasus are two impressive craters, Eudoxus and Aristotles. Note if you can the chalky line extending across the middle of Mare Serenitatis; this is a ray from Tycho, far to the southeast, which is itself still invisible. Right on the ray lies the small but sharply defined crater Bessel.

The region near the central and upper (southern) portions of the terminator at first quarter is the most rugged and tumbled terrain on the entire face of the moon. Craters of all sizes, by the hundreds and thousands, lie cheek by jowl and even heaped on top of one another. Some, like the giants Albategnius and Hipparchus, have been so eroded by millions of years of bombardment that their walls are hard to distinguish among the debris and smaller craters among which they lie; others have outlines seemingly as sharp as the day they were blasted out of the lunar rocks. This is a good time and place to note the actual movement of the shadow line over a period of several hours—the slow illumination, first of the topmost peaks of a crater wall plus perhaps the central peak, then of a crescent on its eastern floor, with the rest in black shadow, and finally the entire floor illuminated, barring a shadow at the base of the western ring wall.

A couple of days past first quarter, when the moon is nine to ten days old, both the Caucasus and the Apennines will be clearly visible to the north, while still farther north, along the northern "shore" of Mare Imbrium, are the Alps. This range is cut by the Alpine Valley, which even more than the Rheita Valley appears to be a graze gouged out by a low-trajectory meteor. Alternatively, it may have been caused by an enor-

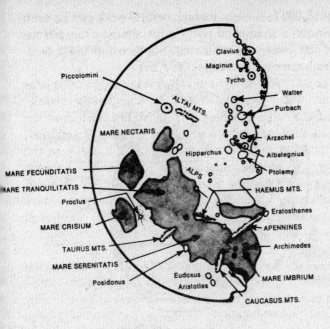

Gibbous moon, about 9¼ days old.

mous chunk of debris from the meteoric super-explosion that (perhaps) formed Mare Imbrium.

To the south, on the other side of Mare Serenitatis in the area between Mare Tranquilitatis and Mare Vaporum, note the curious clefts—some centering on the crater Triensiecker, one extending on both sides of the tiny crater Hyginus, while the largest, Ariadneus' Cleft, stretches almost all the way from Hyginus' cleft to the shores of Mare Tranquilitatis. What produced these Grand Canyon-size cracks is unknown; they somewhat resemble the cracks in the earth left after an earthquake—but no earthquake (fortunately!) was ever powerful enough to produce scars more than a fraction the size of these.

Farther to the south, one sees more of the same tumbled terrain seen in the same general area at first quarter. Note especially the line of six giant craters, beginning with Ptole-

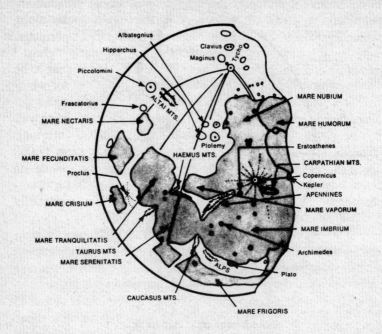

Gibbous moon, about 11¾ days old.

my, almost in the center of the disk, and ending with Walter to the south. Southeast of Walter, Tycho is clearly visible—though its rays will not show up so clearly as later in the month—while north of it is the 110-mile giant crater Maginus and still farther north the moon's largest crater, Clavius, 140 miles in diameter and with an area equal to Massachusetts and New Hampshire put together. Its battered walls still rise to 12,000 feet above its central plain, and in a few places reach 17,000; their original height may have been near 20,000.

At 11 or 12 days, the moon's most conspicuous feature is the great crater Copernicus, whose system of rays streaks much of Mare Imbrium to the north and Mare Nubium to the south. To its left (west) is the hardly less conspicuous crater Eratosthenes; between them are several dozen tiny pits, thought to have been formed long ago by some volcanic process. Many of the pits are arranged in lines, presumably representing areas of weakness in the moon rocks along which many pits could form. Above (south of) Copernicus, note on the dark floor of Mare Nubium the many "ghost" craters—ring walls, or portions of ring walls, that appear to be the remains of older craters submerged by the lava (or whatever it was) that covers Mare Nubium. What may be an earlier stage in this process can be seen in the crater Plato, lying well to the northwest on the opposite shore of Mare Imbrium. Though its walls are fairly well defined, they are low (7,000 feet) in comparison with its diameter (60 miles), while its floor, like that of a mare, is dark—one of the darkest spots on the moon, in fact—suggesting that the crater was partially filled by molten material.

In another day or so, you will be able to see Kepler, almost due east of Copernicus, while north of Kepler is a third rayed crater, Aristarchus, forming almost a perfect right triangle with the other two (Kepler is at the right angle). This group is, of course, even more striking at full moon (between 14 and 15 days after new moon), in which the rays—along with those from Tycho, Stevinus and the others we have mentioned —will be at their brightest.

In the period between full moon and new moon, the various features we have been describing will disappear in the same order they appeared, with the terminator appearing in the moon's "far west" and gradually engulfing the disk over a period of two weeks. There is little point detailing the moon's appearance during this time—first because it is only visible after midnight, an inconvenient time for most people, or, more dimly, during the day, but mostly because we would simply be repeating the past several pages. However, it is worth staying up late (or getting up early) at least a few times during the moon's waning fortnight, to see the features you are already familiar with illuminated from the opposite direction—which in some cases will bring out new details. But there is little reason to bother with this until you are really familiar with their appearance during the fortnight *before* full moon—which will take several months of observations.

Chapter 7

NEIGHBORS IN SPACE
Planets and Other Nearby Sights

Unlike the stars and (to a lesser extent) the moon, there is very little about the planets that is interesting to look at with the naked eye, or even with binoculars. For this reason, I have not devoted a separate chapter to the planets seen without a telescope—it would almost resemble that famous chapter from an old work on natural history, concerning the snakes of Ireland, whose entire text read "There are no snakes in Ireland." Readers who lack a telescope will probably still want to read this chapter, since even though they will find the planets of minor visual interest, it's still fun to know where they are and what they look like.

The first thing that can be said about the planets is that (as almost everyone knows) they are far closer to us than the stars—though much more distant than the moon. Stellar distances are, as we know, measured in tens, or hundreds, or even thousands of light-years, while planetary distances are measured in light-minutes or, at most, light-hours. (Much more usually, of course, they are expressed in millions or billions of miles or in astronomical units—see Glossary.)

The second thing about planets is that their positions are constantly changing relative to the stars—though of course some change more rapidly than others. The stars, as we have already noted, are "fixed," meaning that their positions relative to one another do not change noticeably in a lifetime or even several lifetimes. The planets, by contrast, move through the heaven in a way that led the Greeks to christen them *planetes,* "wanderers." Their wanderings, however, are confined to a relatively narrow belt, near the Ecliptic—the imaginary line that marks the sun's path through the heavens. The constellations along this line are virtually identical with those of the zodiac; the exception is Ophiuchus, which shares with the zodiacal sign Scorpius the portion of the Ecliptic between Libra and Sagittarius. There are all sorts of intricate methods for calculating where a given planet will be in a given month of a given year, but for simplicity I have merely put together a table giving the constellations in which they will be found during the summer months for the years 1981 to 1988. So far as the four "high-visibility" planets are concerned, once you know the region of the Ecliptic where the planet is, you should have little, if any, difficulty recognizing it.* The first step in planet-finding, therefore, is knowing how to locate the various zodiacal signs. (The Ecliptic is also marked in the four monthly star charts.)

Gemini, the westernmost of the summer zodiacal constellations, is easily found in early summer (in late summer it sets before nightfall) by its twin stars, Castor and Pollux (see page 31); the Ecliptic passes some 6° south of the latter star. Cancer, just to the east, is (as noted earlier) hard to distinguish. If, however, you can locate the Beehive cluster, the most distinctive object in that constellation (see page 60), you can get a line on the Ecliptic by noting the fourth-magnitude star about a degree below the Beehive and to its left—the most conspicuous star in the area, though that is not saying much; it lies almost exactly on the Ecliptic.

Moving on into Leo, the Ecliptic passes just south of the constellation's dominating star, Regulus, and in Virgo, still farther to the east, a couple of degrees north of *its* dominating star, Spica.

Libra is a hard constellation to find, since its brightest star is not quite second magnitude, while its next brightest is no

---

*Anyone needing more detailed information on planetary positions will find it, month by month, in the magazines *Sky and Telescope, Astronomy* and *Natural History.*

more than third. I have a certain affection for these two stars, however, because of their names, which sound like incantations out of the Arabian Nights: Zubeneschamali and Zubenelgenubi (Zuben'ubi for short). The latter is the one we are interested in, since the Ecliptic runs just below it: It can be located almost midway between Spica and Beta-Sco, the uppermost of Scorpius' three "claw" stars.

Scorpius itself can be located easily, of course (see page 30); the Ecliptic passes between the upper two of the three claw stars. So far as Ophiuchus is concerned, the constellation is not too difficult to identify—but not very helpful in locating the Ecliptic, which runs well south of its more conspicuous stars. It makes more sense, perhaps, to think of the "Ophiuchus" part of the Ecliptic as simply lying between Scorpius and Sagittarius, both of them easily identified. Thinking of Sagittarius as "the two bows and arrows" (see page 38), the Ecliptic runs just north of the top of the first bow, and just south of the top of the second.

The next three zodiacal signs, Capricorn (technically, Capricornus), Aquarius and Pisces, are all real dogs to identify. Capricorn is marked by widely scattered third-, fourth- and fifth-magnitude stars, which do not group into any very striking patterns. Perhaps the simplest method, though only an approximate one, is to use Vega and Altair as pointers: The path of the Ecliptic through Capricorn lies about as far from Altair on one side as Vega does on the other.

Aquarius is even tougher to locate. Run your eye along an imaginary line from Vega through Gamma-Del, the "nose" of the Dolphin (see page 35) and continue for a slightly greater distance; this should place you near the Ecliptic in Aquarius. In Pisces the Ecliptic can be found by a similarly approximate method, using one side of the Great Square as a pointer. Start with Alpheratz, which lies, as you may recall, on the line from Polaris through Caph in Cassiopeia; continuing along almost the same line brings you to Algenib. The Ecliptic in Pisces lies on the opposite side of this star from Alpheratz, and about the same distance away.

Aries, our next-to-last zodiacal sign, is somewhat easier to locate, since it contains the second-magnitude star Hamal, made more conspicuous by the fact that it is the only star of comparable brightness in its immediate neighborhood (a third-magnitude star lies close to it). Here your pointers are Beta-Peg and Alpheratz—the side of the Great Square *facing* the Milky Way; following this line will bring you to Hamal and, continuing another 12° or so, to the Ecliptic.

Our final zodiacal sign, Taurus, is of course unmistakable, marked as it is by Aldebaran and the Pleiades (see page 77); the Ecliptic runs halfway between them.

With these directions, plus the monthly star diagrams, you should be able to locate the general area of the Ecliptic in any given sign without too much difficulty; the following table will then tell you in what sign to look for a given planet in the summer of that year.

Note that the sign in which a planet lies tells you not only where to look for it but also when. Roughly speaking, Cancer and Leo are visible only in early summer, in the evening, while Virgo and Libra are visible in late evening in early summer, but earlier as the season advances. Scorpius and Sagittarius are visible fairly late in early summer, gradually becoming earlier until by September neither can be seen at all. Capricorn, Aquarius, Pisces and Aries are visible for at least part of the night during most or all of the summer, while Taurus and Gemini can be seen only in the hours after midnight, in late summer.

Before going into what you can expect to see when you look at the planets, let us first note the two main categories into which they are divided: "inferior" and "superior." The inferior planets, Venus and Mercury, have orbits lying between the earth and the sun—*i.e.*, they are always closer to the sun than we are. This means, first, that they are visible only as "morning" and "evening" stars—fairly close to either sunrise or sunset; you will never see either of them at midnight or the hours immediately before or after. Second, the inferior planets, because of their position, show phases somewhat like those of the moon, growing from a thin crescent to a half-illuminated disk ("full planet," corresponding

TABLE OF SUMMER PLANETARY POSITIONS

VENUS

	Early (June 1)	Late (Sept. 1)
1981	[Taurus]	Virgo (P.M.)
1982	Aries (A.M.)	Cancer (A.M.)
1983	Gemini (P.M.)	[Leo]
1984	[Taurus]	Leo (P.M.)
1985	Pisces (A.M.)	Cancer (A.M.)
1986	Gemini (P.M.)	Cancer (A.M.)
1986	Gemini (P.M.)	Virgo (P.M.)
1987	Aries (A.M.)	[Leo]
1988	Gemini (P.M.)	Gemini (A.M.)

	MARS	JUPITER	SATURN
1981	[Tau-Gem]-Cnc (A.M.)	Virgo	Virgo
1982	Vir-Lib (P.M.)	Virgo	Virgo
1983	[Tau-Gem-Cnc]	Scorpius	Virgo
1984	Lib-Sco (P.M.)	Sagittarius*	Libra
1985	[Tau-Gem-Cnc]	Capricornus*	Libra
1986	Sag*	Pisces	Scorpius
1987	Gem (P.M.)-[Cnc-Leo]	Pisces	Ophiuchus
1988	Aqr-Psc (A.M.)	Taurus	Sagittarius*

A.M. = morning star (visible only or mainly after midnight)

P.M. = evening star (visible only or mainly before midnight)

[　] = invisible

*Opposition—maximum brightness and visibility—occurs between June 1 and September 30.

Note: The fact that a planet is not listed as "invisible" does not imply that it will necessarily be visible the entire summer or the entire night.

to full moon, occurs when they are on the opposite side of the sun, hence invisible).

The superior planets—all the rest of them—lie, as you might guess, *farther* from the sun than we do, so that they may be present in the heavens not just as morning or evening stars but at any time. Moreover, they do not show anything really comparable to the phases of the inferior planets— though Mars, at least, sometimes appears somewhat flattened on the one side or the other, owing to its disk's not being completely illuminated.

As most people know, planets, unlike stars, shine only by reflected light; they are not, as the astronomers say, self-luminous. (The same is, of course, true of the moon.) Despite this fact, Venus—the most conspicuous planet, and the only inferior planet you are likely to see much of—is far brighter than any star; its magnitude ranges from −3.3 to a peak of −4.4. When visible, it will appear to the naked eye as a super-brilliant yellowish-white star in the west, for an hour or two after sunset, or in the east an hour or two before sunrise. Because of its brightness, Venus becomes visible almost immediately after sunset and sometimes even during the day, provided you know precisely where to look for it. Binoculars reveal the most noticeable difference between stars and planets: the latter do not appear as brilliant pinpoints of light but show either a disk (in the case of the superior planets) or the phases mentioned earlier. Whether you can actually see the phases of Venus in binoculars depends principally on how steady you can hold them, but in any case it won't look like a star. Even a small telescope, however, will clearly show the phases, though nothing else; Venus is covered by a dense mantle of clouds which even space probes can make little of.

Mercury, being both smaller and farther from the earth, is less bright—its magnitude ranges from −2 to +1—and far harder to see; since it never strays far from the sun, its light is

Venus, like the moon, shows phases, but is never seen much more than half illuminated, since at such times it is either behind or very close to the sun. Mercury shows similar phases; other planets do not.

generally swamped by the glare or sky glow from that body. A few times a year, however, it can be seen as a tiny orange-white "star" for an hour or so before sunrise or after sunset; the best way of locating it is not by the sign it is in—when it is visible, the constellations will not be—but by consulting the current issue of one of the magazines mentioned earlier. A telescope will show its phases—but work fast, or it may be gone before you get it in focus! The phases, alas, are all you will see; even observatory telescopes show little of the planet's surface features. It took space probes to reveal that the surface of Mercury is cratered rather like that of the moon.

Both the inferior planets are most easily visible at what is technically called their greatest elongation—when the angle between them and the sun (as seen from earth) is greatest. Venus can achieve an elongation of as much as 48°, while Mercury's is never more than 23°; if you can visualize the size of this angle (about one quarter of a right angle) you will understand clearly why Mercury is so hard to spot.

The same principle—the greater the angle between planet and sun, the better the looking—also applies to the superior planets, though the term "elongation" is for some reason not applied to them. Since they lie outside the earth's orbit, their maximum angle is 180°—*i.e.*, the planet lies on precisely the opposite side of the earth from the sun—at which times they are said to be in opposition. Since opposition is also the time that the planet is closest to earth, the period around this time is naturally the best time for observing it. By contrast, when the planet is at or near conjunction—lying on the opposite side of the sun from earth—it is invisible.

Mars revolves around the sun in something under two years, so that its oppositions occur at about that period, with conjunction occurring in the alternate years. Because its elliptical orbit is relatively eccentric—elongated—its distance from earth even at opposition (and therefore, of course, its brightness) varies considerably, from 35 million to 42 million miles. When near opposition, it appears in the night sky as a brilliant, reddish body, with a magnitude as much as –2.8 at the "best" (closest) oppositions, and even when unfavorably placed—*i.e.*, when it rises at dawn or sets at nightfall—it is still noticeable, with a magnitude of +1.8.

Binoculars held steady will show Mars as a disk—larger or smaller, of course, depending on its distance—but no more than that. Indeed even a 3-inch telescope will show little in the way of surface detail, except under optimum conditions. In 1969, I was lucky enough to be able to observe Mars through my 2.4-inch instrument at one of its closest oppositions in this century. Blazing red in the midnight sky, it was an impressive object indeed—but my attempts to see more, by running up the magnification, merely blurred its outline, along with any surface features. I did not then have my 6-inch instrument, through which I could have seen some of the planet's dark surface markings (their nature is still disputed) and one or perhaps both of its brilliant white polar caps (thought to be thin deposits of ice or hoarfrost). Mars has two tiny moons, Deimos and Phobos, but don't bother looking for them outside an observatory.

Jupiter is—to my taste at least—the most interesting of all the planets. For one thing, it is always noticeable (if it is visible at all); even at its dimmest it is bright as Sirius, and at opposition its yellowish-white light blazes forth at magnitude −2.5, close to the maximum brightness of Mars. Jupiter, moreover, possesses surface markings visible through most telescopes. Even through my 2.4-inch refractor, I can see the suggestion of markings across the disk, and in my 6-inch reflector they show up clearly, even at low magnifications—two tannish bands near the planet's equator, and another, somewhat less noticeable, closer to one of its poles. If the timing is right, the 6-inch scope also shows the "red spot," a blob of color close to the planet's equator whose nature is still a subject of argument among the astronomers.* With the Red Spot in view, one can chart the planet's rotation by the spot's progress across its disk; Jupiter, despite its enormous size (it is

---

*The most recent theory, based on observations from space probes, is that the spot is the top of an enormous hurricane in the Jovian atmosphere. Nobody, however, has yet explained why only one such super-hurricane has ever been observed on the planet, or—even less—how it has continued to exist for more than three centuries, as we know it has. On earth, certainly, the lifetime of hurricanes or other large storms is measured in weeks or even days.

nearly 11 times the diameter of earth), rotates more rapidly than earth, making a complete revolution in less than 10 hours. Thus the Red Spot takes something under 5 hours to pass from one edge of Jupiter's disk to the other, and will show a perceptible shift in position over an hour or two.

Perhaps the most interesting thing of all about Jupiter, however, is its moons, with which it comes better equipped than any other planet. Of its 14 satellites, ten are invisible outside an observatory, but the four Galilean moons* are visible with any telescope and even with binoculars—how many you can see in this way is a test of how steady you can hold them.

Even through the smaller of my two telescopes, the four moons show up clearly as tiny "stars" arranged on either side of the planet. Because they revolve around their parent quite rapidly—much more so than our own moon—their aspect changes from night to night and even from hour to hour: sometimes all four on one side, sometimes three on one and one on the other, or two on either side, at the same time varying their apparent distance from Jupiter. As a result, you are very unlikely ever to see the moons in exactly the same configuration.

Sometimes one of the moons is "missing"; either it has passed in back of its parent and is eclipsed by it or it is passing in front, with its light swallowed up in that of the planet's bril-

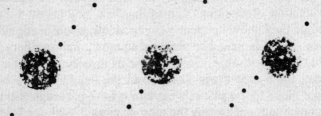

Jupiter and its four major moons provide the stargazer with a constantly changing panorama from night to night. In the middle drawing the "missing" moon is in eclipse, having passed behind the planet.

*So called, of course, from their discovery by Galileo with his pioneering telescope.

liant disk. Occasionally two moons are "missing," and very occasionally, three. Once in a very long while, I suppose, all four moons reach a configuration such that all are invisible, but I have never seen this and don't expect to.

Astronomical magazines often include monthly tables or diagrams, showing the changing arrangement of the moons from night to night during the month, but personally I think it more fun to discover this for yourself. You can begin by simply noting how many of them are visible on a given night, and on which side of Jupiter. Then try making a rough sketch of their position, spend an hour or two looking at other objects (or throw a blanket or tarp over the telescope and do something else) and then look at them again; they should appear noticeably different from your sketch. The differences will be particularly marked, of course, if you happen to catch one of the moons just passing into, or emerging from, invisibility.

When one of the satellites passes in front of the planet, or sometimes even when it is just about to, it casts a shadow; this occurs, on the average, almost one night out of two—though only for a part of the night, since the shadow passes across the disk in 3½ hours or less. They are clearly visible, as

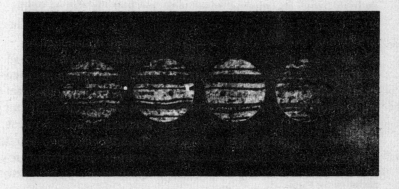

When a moon passes in front of Jupiter, its shadow can often be seen in a telescope passing across the planet's disk. Sometimes the moon itself can also be seen, as a slightly lighter or darker spot on the disk.

tiny, black dots on the face of Jupiter, in a 6-inch telescope, but probably not in a 3-inch model. Tables of these satellite passages are also listed in astronomical magazines—though again I think it's more fun to discover these events for yourself. Look for a night on which only three (or two) satellites are visible, or in which one is very close to the disk; if the missing satellite is passing in front of Jupiter (rather than in back of it) or if the close-in moon is about to pass in front, you will be able to see its shadow (assuming, of course, that you have a 6-inch telescope). Once in a long while—probably no oftener than once a year—you may see *two* shadows simultaneously. Occasionally, if the night is clear and your eyes are good, you may see both the shadow and the satellite, the latter as a slightly lighter or darker spot on Jupiter's disk.

Jupiter takes almost 12 years to revolve around the sun, meaning that—speaking roughly—it moves into a different zodiacal sign every year. During the early 1980s, it will be visible before midnight during all or most of the summer; from 1985 to 1987, it will be visible before midnight only in late summer, and in 1988 will not be seen at all before midnight during the summer months. In 1989, it will be invisible in early summer (later, visible after midnight) and in 1990 will be visible in early summer just after sunset—at which time its moons may well be lost in the skyglow. Thereafter, Jupiter will again become a "summer" constellation.

The last of the "interesting" planets is Saturn—mostly, of course, because of its famous rings, those strange flattened belts of debris (mainly ice, it is thought) which give the planet its unique appearance. Saturn is almost as large as Jupiter, but more than twice as far off, meaning of course that it is less bright. Even at its dimmest, however, it is a respectable first magnitude, and at opposition rises to −0.4—brighter than Arcturus. Through steady binoculars, it appears as a tiny, dullish

---

*Close-in observations from unmanned spacecraft have revealed that Jupiter too has a ring. However, it is too faint to show up even in an observatory telescope. Similar observations have revealed the startling fact that Io, one of the Galilean moons, possesses an active volcano—the only such object known anywhere but on earth.

yellow-white disk, but seeing the rings requires a telescope. One book I have read claims that at least a 3-inch scope is needed for this purpose, but I have seen them with perfect clarity in my 2.4-inch instrument; my 6-inch, of course, gives an even bigger and brighter picture. Saturn is almost as well equipped with moons as Jupiter; of its 11, one or two can be seen with a small scope and—at opposition under the best conditions—as many as five with a 6-inch. Saturn's pace around the sun is even more leisurely than Jupiter's—a full revolution takes nearly 30 years, meaning that it takes some 2½ years to move from one sign to the next. Currently (1981) it is in Virgo, and therefore well situated for observation in early summer. During the balance of the decade it will become visible for more and more of the summer night and will be visible at least in late summer through the 1990s.

Uranus and Neptune, the two remaining large planets, are—to my taste—interesting only to those who like to look at celestial objects "because they're there."* Neither of them has either moons or surface markings visible in any but a large telescope, and both are so inconspicuous as to make them difficult to find. Uranus, at magnitude 5.7, is at the very threshold of naked-eye visibility, though it shows up clearly enough in binoculars—provided you know where to look, and can distinguish it from the surrounding stars. With a 6-inch telescope, you should be able to see it as a tiny, greenish-blue disk. Neptune, at magnitude 7.6, is invisible to the naked eye and dim even in binoculars, making the problem of picking it out from its stellar surroundings correspondingly more difficult; indeed the only way to be certain that you are seeing it is to know its exact position in the sky, relative to nearby stars—and then to know at precisely which of these stars your telescope is pointing. As it happens, both planets are currently well situated for summer observation—Uranus in Libra, Neptune in Sagittarius—and will remain "summer planets" until well into the next century, so leaden is their pace

---

*Pluto, last and most distant of the sun's family, is not even "there" so far as the amateur is concerned. Unless one happens to have access to at least a 12-inch telescope, its fourteenth magnitude disk is invisible.

around the sun. If you feel driven to try and spot these two elusive objects, get hold of a recent issue of *Astronomy* or *Sky and Telescope,* or a copy of the current *American Ephemeris and Nautical Almanac,* any of which will tell you the planets' precise position. Then say a small prayer to whatever god you favor—and start looking. Lots of luck!

The Sun

The first thing that should be said about observing the sun is—don't! Not, at least, until you know *exactly* what you are about—and even then, with the utmost caution. Merely looking at the sun with the naked eye can blind you temporarily— or permanently, if you are crazy enough to do it long enough; binoculars or a telescope, used without the proper precautions, merely speed up the process. Galileo is said to have damaged his eyes in this manner.

The basic problem in solar observation, of course, is to cut down the sun's brightness sufficiently so that it won't damage your eyes—even when concentrated. With binoculars, there is only one simple way to do this, and it requires fairly special weather: considerable atmospheric haze, but little or no cloud. In this situation, you can observe the sun near its setting under the following conditions. READ THEM CARE-FULLY!

1. The sun must be within a few degrees of the western horizon.

2. It must have taken on a red color—not orange, but *red.*

3. You must be able to look at it with the naked eye without squinting—in fact without even *wanting* to squint.

If, *and only if,* these conditions have been met, try a cautious peek through your binoculars; if your eyes are still quite comfortable, try a longer peek. Using the atmosphere in this way as a filter, you can observe sunspots, which will appear on the red disk as black flecks of various sizes and shapes. Almost invariably you will see at least one, and at periods of sunspot maximum—which occur about every 11 years (no one knows why)—you may be able to see dozens. What you are seeing are "storms" in the sun's atmosphere—areas in which its flaming gases are rising, therefore cooling, therefore emitting less light. They are not, in fact, really black, merely less bright than their surroundings; it is the filtering effect of the atmosphere that makes them look black.

If you are lucky enough to get several days running of hazy weather, you can mark each evening the progress of a particular spot, or group of spots, as it moves across the sun's disk —proving that the sun rotates on its axis, just as the planets do. Oddly enough its equatorial regions move faster than its polar regions—about 25 days for the former as against 30 days for the latter.

To observe the sun through a telescope you will need special equipment. Perhaps the safest method is to use what is called a sunscreen—a flat, white-painted piece of metal which can be mounted on your scope in such a way that the sun's image can be projected onto it through the eyepiece. Aiming the telescope can be tricky, however, since *you cannot safely use your spotter scope without additional precautions*—for example, a piece of black, overexposed film negative held in front of it. Even then, use dark glasses, and take a quick, trial peek first. As a general rule; and regardless of whatever precautions you may be taking, *if your eyes are the least bit uncomfortable in looking at the sun, you shouldn't be looking at it.* Alternatively, you can simply point the scope in the general direction of the sun and move it about until it casts the smallest possible shadow. At this point, a little juggling should get the solar image properly centered on the screen.

Another method, usable with telescopes up to 3 inches, involves the use of a sun cap, a plastic cap with a hole in the middle. Fitted over the "mouth" of the telescope tube, it narrows the diameter of the mouth, thereby cutting out a major proportion of the incoming light. A sun cap, however, must always be used with a sun filter—a disk of almost pure black glass that screws into the eyepiece. This method is unsafe on telescopes larger than 3 inches, which will concentrate enough light so that the eyepiece filter may crack.* Another disadvantage is that the sun cap, by narrowing the effective "aperture" of the scope, will cut down the amount of detail you can see, which as already explained depends on the aperture.

Yet another method, usable on telescopes of any size, involves purchasing a "sun prism"—a specially ground piece of glass mounted in a short tube that is inserted between the telescope tube and the eyepiece. The prism is ground at angles such that the beam of sunlight is reflected five times before passing into the eyepiece, losing much of its intensity at each reflection; the net result is that only about one percent of the light reaches the eyepiece, the remainder being absorbed in the body of the device. Though safe, sun prisms are fairly expensive; they can be gotten through many scientific supply houses.

Once you have taken the proper precautions, you can then use your telescope to observe not merely sunspots but some other solar features—notably, "flares"—eruptions in the sun's atmosphere which send giant plumes of gas hundreds or even thousands of miles aloft. Flares are uncommon, but are interesting if you are lucky enough to spot one; they show up as bright flashes, usually near a sunspot.

Special Events

Some years ago, not long after I started stargazing, I was watching the stars come out one evening when I noted that the Big Dipper had acquired an eighth star. I blinked, and for a moment wondered if I hadn't spent too much time in the sun that afternoon. Then, putting my binoculars on the stranger, I realized that it was slowly moving. This was my first observation of an artificial earth satellite.

The most likely time to see a satellite is after sunset, about

---

*Do not under any circumstances use a filter in or over the eyepiece that is made of celluloid or any other plastic; it will almost certainly melt or catch fire.

the time the stars are appearing. The sky will already be fairly dark, while any satellite in the neighborhood, being several hundred miles (or more) up, will still be reflecting the sun's rays. (If it is high enough, of course, it will reflect them for considerably longer; I have observed satellites around 10:30 P.M., more than two hours after sundown.) A moving "star" in the sky is either a satellite or a high (or distant) plane; binoculars will settle the matter, since the plane will be carrying a red light (often flashing) as well as a white one; satellites are the yellow-white of the sunlight they reflect, and of course do not flash. Not infrequently, however, they can be seen to wax and wane regularly in brightness, within a period of several seconds. This shows that the satellite is "tumbling"—turning end for end, in such a way that the surfaces pointing toward you are sometimes more reflective, sometimes less.

Time was when some newspapers printed lists of satellites (usually in the columns devoted to weather maps and forecasts), giving the time they would pass overhead, and in which portion of the sky they could be seen. Nowadays, however, the vicinity of the earth is so cluttered up with communications satellites, scientific satellites, spy satellites and I don't know what else that the papers have given up trying to keep track of them, so that you must depend on luck to spot one. However, in a summer or two of stargazing, you are likely enough to see at least one or two.

Another kind of moving star is, of course, the "shooting star," which as almost everyone knows is not a star at all but a meteor—a bit of cosmic debris which has hit the earth's atmosphere. Friction with the atmosphere first heats and then gasifies the meteor, which thereby leaves a brief trail of hot, luminous gas in the upper atmosphere. Most meteors are produced by particles ranging in size from a dust particle to a grain of sand. The larger the particle, of course, the longer it takes to "burn up" in the atmosphere—and the longer and brighter its trail. Perhaps one meteor in a million is big enough to reach the earth's surface before having lost all its substance. It is then a meteor*ite*—a chunk of rock or nickel-iron whose pitted surface shows the effects of its fiery passage through the atmosphere. When a really large meteor hits the earth, its impact can create an explosion larger than that of an atomic bomb. Meteor Crater in Arizona is the scar left by just such an explosion, perhaps 50,000 years ago, and aerial photographs have revealed similar scars in many other places —some of them dating back several hundred million years. After that lapse of time, of course, they no longer look like craters, but traces are still there for geologists who know how to spot and interpret them.

An hour's sky-watching on almost any clear, moonless night will produce one or two meteors, and you will miss as

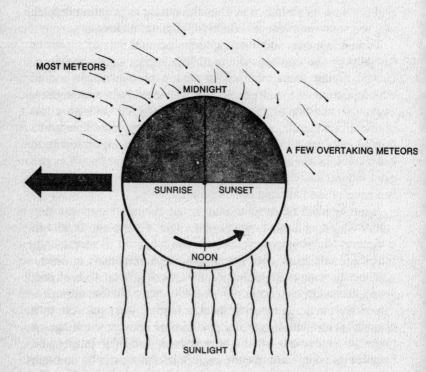

The hours between midnight and dawn are best for observing meteors ("shooting stars"); this schematic drawing shows why they are more plentiful then.

many more because you are looking at some other part of the sky when they hit. In the hours after midnight (*i.e.*, one A.M. Daylight Time), the rate rises to as much as 10 per hour. The reason for the difference is that in the later hours the part of the earth where you are standing faces the direction the earth is moving in its orbit around the sun, and thus "sweeps up" any bits of cosmic junk that happen to be in the neighborhood. Earlier in the evening, however, you will be facing backward into the already "swept" area, meaning that the only meteors you are likely to see are those moving fast enough to overtake the earth. Thus the "ideal" time to spot meteors is between 1 A.M. (D.T.) and dawn, during the period between new moon and first quarter, when there will be no moonlight during the early morning hours. Second (very definitely second) best is the hours before midnight during the moon's last quarter.

At certain times of the year, the earth passes through a particularly dense aggregation of debris, and the rain of meteors increases sharply, to as much as 50 an hour after midnight and perhaps 15 an hour even before it. One of the largest and most regular of these meteor showers occurs in mid-August; it peaks during the night of August 11–12 or 12–13, but is noticeable for a couple of days on either side of the peak date. If the moon is "right" on one of those dates, don't miss it, even if you have trouble staying awake.

An interesting project during this shower—assuming you already know your constellations pretty well—is to take outdoors with you a small chart of the heavens, a pencil or pen and a small flashlight—preferably with a red bulb or filter, so as not to temporarily destroy your night vision. As each meteor streaks across the sky, try to note its approximate path and pencil it in on the chart. After a while, you will notice that most of the tracks appear to converge on a point in the constellation Perseus—for which reason this August shower is called the Perseids. Don't expect all the tracks to converge, however; even in the midst of the Perseids you will still be seeing the occasional random meteor, of the sort that shows up any night.

Unfortunately, the only other notable meteor showers that can be predicted with any certainty occur in months which in most places do not lend themselves to prolonged sky-watching. The Orionids (from their apparent point of origin, or "radiant," in Orion) come on October 20, the Leonids on November 16, the Geminids on December 13 and the Quadrantids on January 3. The Aquarid shower occurs regularly on July 29, but the maximum rate is only about 20 an hour after midnight—about double the normal fall.

Many other showers occur—for unknown reasons—irregularly, and very occasionally produce really spectacular displays. Such was the Bielid shower of November 14, 1827, in which more than 5,000 meteors fell per hour; it has come down in American folklore as "the night the stars fell." Still more remarkable was the Leonid display of 1883, in which the hourly total reached 10,000—far above the usual number. The most spectacular display on record, however was the Giacobinid shower of October 9, 1933, in which the heavens rained "stars" at the rate of some 20,000 per hour, with up to a dozen trails visible simultaneously. If you are lucky, you may see a shower of this sort some time during your life—but don't hold your breath. Astronomical magazines, and sometimes even the daily press, will often clue you in on when an unusually dense shower *may* occur—but these predictions, unfortunately, are not very reliable.

Comets are another type of celestial "special event." Ordinarily, your chances of seeing a really bright one are not much better than those of seeing a really spectacular meteor shower; it happens, however, that early 1986 will bring a visit from one of the most conspicuous—and certainly the best known—of these bodies: Halley's comet.

Despite its name, this comet was not discovered by the distinguished English astronomer Edmund Halley (1657–1742); one might as well talk of discovering the Eiffel Tower or the Brooklyn Bridge. What Halley did was observe the comet in 1682, calculate its orbit (using the just-published Law of Gravitation of his even more distinguished friend,

Isaac Newton) and declare that it must be the same comet that had previously appeared in 1607 and 1531. He predicted that it would reappear in 1758, which it did, though he himself did not live to see it. Subsequent research has traced the comet's visits back to 87 B.C. and, less certainly, for more than a thousand years before that (ancient astronomical records are generally ambiguous and often inaccurate). Since 87 B.C., it has appeared 26 times, at intervals of between 76 and 77 years.

From ancient times, comets have been taken as foreshadowing dire events; thus the 1066 appearance of Halley's comet was taken as predicting the successful Norman Conquest of England. Or so, at least, people said afterward. Its most recent appearance, in 1910, inspired H. G. Wells' famous story "The Day of the Comet," in which poisonous gases in the "tail" of a comet threatened to destroy life on earth. In fact, Halley's comet, like all such bodies, is a ball of cosmic junk—mostly ice mixed with ammonia and methane, but with pebbles and mineral dust embedded in it. The tail (seen in some but not all comets) develops when the body moves close enough to the sun so that solar radiation first melts and then gasifies some of this ice; radiation pressure blasts the gas, mixed with fine dust, into a cloud extending *away* from the sun (*i.e.,* as the comet approaches the sun, the tail trails behind it; as it recedes from the sun, the tail leads it). The tail is made visible by reflected sunlight precisely as diffuse nebulae are made visible by reflected starlight (see page 68); its fibrous or hairy appearance gives comets their name, from the Greek *aster cometes,* "longhaired star." Since comets' tails, like comets themselves, contain the poisonous gases methane and ammonia, Wells' tale had a slender factual foundation. In truth, however, they are so rarefied that even were the earth to pass right through a tail, nobody would be inconvenienced.

"Periodic" comets like Halley's—the first to be identified as a single body that reappeared at regular intervals—evidently lose a certain amount of their gaseous constituents at each pass around the sun. In fact, astronomers estimate that within a few hundred passes—a few thousand years—at most the gases of any comet will have boiled away completely, leaving only a small (and invisible) cloud of rocky debris. This destructive process has presumably been going on since the formation of the solar system some 4½ billion years ago, yet we continue to see comets; the supply is evidently being renewed in some manner. In fact there is thought to be a sort of "reservoir" containing billions of them in the outer reaches of the solar system; most occupy stable orbits so far out that we never see them. Occasionally, however, the gravity of the giant planets (Jupiter, Saturn, Uranus and Neptune) will kick one of these far-out bodies into a new orbit that brings it close enough to the sun for us to spot it—and for the sun's own gravity to further influence its motions. Depending on the nature of the resulting orbit, the comet may thereafter return periodically at intervals ranging from a few years to a few decades, or be flipped out of the solar system entirely.

Halley's comet will reach its closest point to the sun (perihelion) on February 9, 1986; at this point it will be well inside earth's orbit—and will, alas, be invisible from earth because the sun will lie between. However, it will be visible to the naked eye for several weeks of its inward and outward journey, and can be seen through binoculars for even longer. Even at its most visible, binoculars will reveal additional details of its structure, while a telescope will show the "fine structure" of the tail, which can be expected to change from day to day and even from hour to hour. The tail as a whole will, of course, enlarge as the comet approaches the sun and then diminish as it recedes.

Regrettably, this will not be one of the "best" appearances of Halley's comet, mainly because, as noted above, at perihelion—*i.e.,* when its tail is largest and brightest—it will be invisible from earth. In addition, the comet is thought to be almost literally running out of gas, by the process of attrition described earlier. And finally—to strike a practical note—unless you live in the "Sunbelt," late fall and winter are not the most pleasant time for sky-watching; the nights will be long and often brilliantly clear—but will range from chilly to frigid. Nonetheless, Halley's comet is worth getting out your warmest woollies for: no equally fascinating celestial event is likely during your lifetime (its next appearance, in 2062, will be seen only by the youngest of us).

As partial compensation for the limitations on earthbound observations of the comet in 1986, we should be seeing some spectacular photographic closeups of it taken from spacecraft. In addition, NASA had planned a "slow" flyby of two comets, Halley's and the smaller Tempel 2, by a special spacecraft equipped not only to photograph it but also to take samples of the gases in the tail and possibly to land on the surface of Tempel's small nucleus. These observations would be of extraordinary scientific interest, in that they would be sampling what are thought to be remnants of the primordial matter that formed the solar system. In the planets, and of course the sun, the composition of this matter has been radically changed by billions of years of heat, pressure and meteorite bombardment; in comets it has been, as it were, preserved in the deep-freeze of outer space. At this writing, unfortunately, Congress has provided no funds for this second mission—according to some accounts, because the hardware required has no military applications—meaning that we may lose a priceless opportunity to learn more about the nature of the system we inhabit.

Though very few comets are as conspicuous as Halley's, in most years there are at least one or two unobtrusive ones. Some are visible to the naked eye—if you know where to look—while others show up only in binoculars or a low-power scope. A subscription to the magazine *Sky and Telescope* will keep you up to date on what comets are visible, and where to look for them.

The general similarity of the mineral constituents of comets and meteorites may suggest to you that the latter are in fact bits of the former; they are. Many meteor showers are "associated" with known comets, meaning that they occur at points in the earth's orbit through which a comet has previously passed. Others, though not associated with comets, doubtless represent debris left by now "extinct" comets —those whose gaseous constituents were evaporated and blown away thousands or millions of years ago. Not all comets produce meteors, however; some (like Tempel 2) never reach the orbit of earth and others (like Halley's) pass well above or below our orbit on their journey to and from the sun.

The final "special event" which you are likely to see during your sky-watching actually has very little to do with astronomy, since it occurs in the earth's upper atmosphere, not in space: the Aurora Borealis ("dawn of the north") or Northern Lights. Just a few months ago, when I was working on the early chapters of this book, I was gazing toward the west one clear evening when I saw a strange, luminous cloud near Arcturus. Within a minute, it had elongated and moved toward the northwest. As my eye followed it, I realized that something extraordinary was going on in that part of the heavens. Shimmering curtains of bluish-green lay along the horizon, at times stretching all the way from northeast to due west. From time to time, narrow, luminous spears would shoot up well above Polaris—45° and 50° above the horizon; at other times the curtains would shimmer like masses of glowing gauze fanned by a breeze. The display continued for several hours— it was, as I learned later, the most intense summer aurora for some ten years.

I have seen the Northern Lights some half a dozen times over the years—mostly in the summer. They are, in fact, no less common in that season than in the winter when many people "expect" them; the difference is that you are most likely to see them in winter because the nights are longer and the air is colder, and therefore drier and clearer. Assuming, of course, that you happen to be out of doors! As most people know, they are caused when the sun sends off a particularly intense blast of subatomic particles which, reaching the earth, excites the atoms of the tenuous upper atmosphere into luminescence. Since the particles in question interact with the earth's magnetic field, they tend to "fall" most thickly above the earth's magnetic poles—in the Northern Hemisphere, a region in northern Canada. Thus the nearer you are to that area, the more likely you are to see an aurora; in northern Canada and Alaska, indeed, auroral displays are almost routine. Even in the northern U.S., however, they are likely to occur once a summer, and oftener in periods of sunspot maximums—sunspots seem to have something to do with them. Perhaps once in ten years, the aurora will be extensive enough to be seen in the southern states and even in Mexico.

The aurora is strictly a naked-eye "object"; neither binoculars nor (even less) a telescope will improve the view. But their ghostly radiance, shimmering and rippling along the northern horizon, is something to see. On any clear night, make a point of glancing at the northern horizon before you go to bed; it may take several years, but sooner or later you will be rewarded.

Chapter 8

STARGAZER, WHAT NOW?
Where You Can Go from Here

AS I indicated at the beginning of this book, what I have tried to write is a very simple introduction to astronomy, not a course in it. By the time you have reached this chapter, however—assuming you have done the observations described in it—you may well be hooked on astronomy, to the point where you want to go further. The following, then, are some projects that may appeal to readers of various types.

As I have also noted earlier, I have made no effort to list a comprehensive catalogue of the interesting and/or beautiful objects in the heavens. If you are of a systematic turn of mind, you may enjoy going through the Messier Catalogue in Appendix A to see how many of the objects listed in it you can spot *and identify* with your scope. (I emphasize "identify" because in some cases several rather similar objects —galaxies, for example—lie close enough together so that you may be in some doubt as to which you are looking at.) Nor does the Messier list exhaust the list of "small telescope" objects; there are a sizable number of quite bright clusters, for example, which Messier for some reason never listed. And there are dozens of attractive doubles I have not bothered to mention.

Perhaps the best guide to these "neglected" objects is Donald H. Menzel's *A Field Guide to the Stars and Planets*, which contains extensive tables listing the brighter celestial objects, as well as a comprehensive atlas of the heavens—a series of charts showing sections of the sky, each facing a photograph of the same area, with descriptive text beneath describing the high-spots of the region in question. Unfortunately, Menzel is a former director of the Harvard Observatory, so that his descriptions of the objects themselves often owe more to high-magnification observatory photographs than to their actual appearance in a small scope. Another of his series of charts gives similar details of the moon's surface, section by section—and here, fortunately, the descriptions match the amateur's view pretty closely.

The *Field Guide* also has sky maps, showing the changing heavens month by month, but I have never found these very useful, since each shows only half the sky (northern or southern), with constellations around the edges badly distorted. Also included is a certain amount of text, but this you will almost certainly find heavy going; Menzel has got to be one of the world's worst writers.* I have listed some other more or less useful books in the bibliography on page 172.

If you are well and truly hooked on stargazing, you will very likely begin thinking at some point about a larger scope—a 10-inch or 12-inch model, say. Scopes of this size, however, don't come cheap—among other reasons, because the demand is so small that each one must almost be custom-made. If you are an avid do-it-yourselfer, are reasonably handy with tools, and possess a basement or garage workshop, you can save yourself a good deal of money by building your own scope—in particular, grinding your own objective mirror, which represents a large part of the cost of any large scope. Mirror grinding, I am told, does not (as you might think) require highly sophisticated skills; it does require enormous patience and a willingness to take pains. Some of the books in the bibliography will tell you how to go about it, while many local planetariums give courses in the subject.†

---

*The same, alas, goes for his book *Astronomy* (see bibliography), which is full of fascinating lore—and stunning photos—but in many respects might as well be written in Chinese.

†If, however, you want to move up merely from a 3-inch to a 6-inch scope, my advice—unless you are an absolutely avid home-workshop devotee—is to buy it, new, secondhand or in kit form. You will spend little (if at all) more, and save a great deal of time and trouble.

A number of amateur astronomers have gotten interested in variable stars. This is a subject I have deliberately neglected, since most of the really interesting variables are dim enough to make locating them difficult for the beginner. (Algol, mentioned on page 74, is a notable exception.) As you grow more familiar with the heavens, however, you may well enjoy tracking down a particular variable (the important ones are listed in the *Field Guide* tables) and logging its variations from night to night or from week to week, as well as noting the difference in the *pattern* of variation between variables of different types (astronomers recognize more than a dozen categories). Amateurs, organized in the American Association of Variable Star Observers, 4 Brattle Street, Cambridge, Massachusetts, have, I am told, done much worthwhile scientific work in this field, using only binoculars or small scopes. The AAVSO publishes a whole series of charts designed to facilitate finding variables and judging their brightness accurately; a stamped, self-addressed envelope will get you details. (Some samples can be found in Menzel's *Field Guide*, pp. 118–123.)

Finally, if you, like millions of Americans, are a camera bug, you have open to you the whole field of sky photography. There is not a great deal I can say about this, since (as I have indicated earlier) cameras and I don't get along. However, several basic points are worth noting.

First, many quite attractive photographs can be made merely with a good camera—*i.e.*, without using a telescope. The main requirement is that the object should be large enough and bright enough to show up with the modest magnification of the camera lens, and with exposures of no more than thirty seconds. At longer exposures, the image will begin to "streak," due to movement of the stars overhead; if, however, your scope has a clock drive, you can perhaps strap or tape your camera to the tube, pointing in the same direction, for longer exposures. In general, however, telescopeless photography is likely to give you best results with relatively bright objects—the moon, of course, and perhaps the brighter planets, as well as the aurora if you are lucky enough to see one.

By contrast, I have seen an amateur photo of the galaxy M-31 which shows less than you can see in binoculars, let alone a scope.

For photographing galaxies, nebulae and clusters, you will need a scope, both to increase the light falling on your film and to enlarge the image. In theory, you could get the same increase in light-gathering power simply by mounting your camera on the scope, as described above, and using very long exposure; in practice, your film would probably end up fogged with stray light from the sky. Photography through a telescope absolutely requires a clock drive, if the exposure is more than a few seconds, since at telescopic enlargements the image will streak almost instantly. You will also require some sort of device, either bought or handmade, for mounting the camera rigidly in the proper relationship to the eyepiece.

In general, I am inclined to think that sky photography should not be undertaken except by people who already understand cameras reasonably well. Which is why I've never gotten into it.

Finally, you may feel like delving more deeply into the inner nature of the objects you have been examining—the processes underlying the birth and death of stars, the nature of the various planets, the structures and evolution of galaxies, even the origin of the universe, insofar as cosmologists are able to speculate about it. An elementary text in astronomy is the obvious answer here; some of the more readable are listed in the bibliography. In addition, check your local university, community college, planetarium and adult education groups, some of which give courses in astronomy open to—and more or less geared to—the general public.

The intellectual side of astronomy is a smorgasbord offering food for almost any taste. One can ponder the diverse nature of the planets—the cloud-covered, superheated deserts of Venus, the chill near-desert of Mars (which may, or may not, contain living organisms of some sort), the thousands-of-miles-deep turbulent atmosphere of mighty Jupiter. One can learn about the slow processes of compaction which form matter into stars, the nuclear reactions that set them alight and keep them blazing for millions or billions of years, their explosive transformation into novas and supernovas, and their final decline into super-dense "white dwarfs," a teaspoonful of whose material would weigh many tons, or the even stranger "black holes" from which neither light nor matter can escape. One can puzzle over the diverse shapes of galaxies—spherical (almost like super-giant star clusters), elliptical, spiral like our own, or the strange "barred spirals," like a pinwheel crossed with a propeller.

You can work through the intricate methods by which astronomers measure distances in space—the surveyor's parallax, which locates the nearer stars, the Cepheid variables that give clues to the distance of globular clusters and nearby galaxies, and the spectroscopic "red shift," which carries our measuring rods to the outermost limits of the observable universe where the mightiest galaxies appear as mere flecks of light on enlarged observatory photographs. You can share with the astronomers the controversy over quasars, which are either impossibly distant galaxies of no known type or nearby objects which behave in ways no astrophysicist can yet comprehend. You can in imagination travel back to the "Big Bang" with which this universe is thought to have begun—or forward to another Big Bang which, say some cosmologists, will after billions of years mark the death of this universe and the beginning of another. Astronomy can carry you from the solid earth on which you stand to the ultimate distances of space, the ultimate past, the ultimate future.

A Note on Equipment

Since I am expert neither in mechanical nor in optical engineering, anything I have to say on equipment for the amateur astronomer should be taken with a grain of salt.

My two pairs of binoculars are a wide-angle 7 × 35 and a "zoom" 7.5–15 × 40. They are adequate for my bird-watching and star-watching needs, which is about all you can expect of any binoculars. A pair similar to the first will cost you about $40 at Sears, while a simpler model, without the wide-angle feature, will cost you under $25 from the same source.* If you plan to use your binoculars for star-watching only, I'd recommend the cheaper model; for other purposes (*e.g.,* bird-watching) the wide-angle feature is worth the extra money. For reasons given in Chapter 3, I do not recommend zoom binoculars at any price; put the extra money away toward a telescope.

My smaller telescope is a 2.4-inch (60-mm) refractor, with equatorial mounting; a comparable scope will cost you $230 at Sears. This does *not* have an electric clock drive, but since the main point of such a scope is portability, I don't consider the omission important.

The cheapest scope you can get is a 3-inch altazimuth mounted reflector from Edmund Scientific (101 East Gloucester Pike, Barrington, N.J., 08007), for $130—a good "first scope" for kids, who can lose interest fast, but not one I would recommend for serious stargazing. A possible compromise on price—and the most portable cheap scope you can buy —is Edmund's "Astroscan," at $190. This gives you 105 mm (over 4 inches) of aperture—more than three times the light-gathering power of a 60-mm instrument. This scope is of the "rich field" type, meaning

*All prices as of fall 1980.

171

essentially that it trades a wider field of vision—about three times that of conventional scopes, even at low power—and brighter images for lower magnification. It does not include a finder scope, which because of the wide field of vision is supposedly unnecessary. Since I have never used this scope, I cannot say how far this lack complicates the task of finding specific objects; I would guess that a certain amount of practice would be needed before you learned to point the scope where you wanted it to go. But if portability is important (the whole thing weighs a little over ten pounds) you should certainly give it serious consideration.

My larger scope is a 6-inch Dynascope reflector made by Criterion Scientific Instruments (620 Oakwood Avenue, West Hartford, Conn., 06110); its current price of $295 (including crating charge) makes it far and away the best buy in this category. You can shave $40 off the price by eliminating the electric clock drive, but I would recommend not doing so if there's any chance you *might* be able to use such a drive now or in the future. It is a real convenience for simple stargazing, and for sky photography an absolute necessity. The price does *not* include slow motion on the declination axis, which costs another $29 as an option. Spend the money. Weighing some 60 pounds, the instrument is only marginally portable, and this goes double for Criterion's 8-inch reflector, which is usable only where you can set it up once a season and leave it —if necessary, covered by a tarp. Its price of $475 puts it well beyond the beginners' class.

If you are really thinking about spending this kind of money, you should consider the Criterion Dynamax-6. This fine catadioptric instrument is compact, portable (15 pounds) and at $625 a bargain compared to comparable scopes. The price includes clock drive, slow motion on declination axis, and two eyepieces (eventually, you will want at least one more). The Dynamax can be set up on any flat, level surface, such as a small, sturdy table (a tripod will cost you $200 extra). An 8-inch version of the same scope sells for $790 and is almost as portable. Both instruments, with the aid of a special attachment, can be used as super-telephoto lenses for earthbound photography.

Note that none of the prices I have quoted includes shipping costs, which will of course depend on where you are located in relation to the manufacturer or dealer.

Note also that eyepieces are not always interchangeable from one scope to another. Specifically, Japanese scopes such as the 2.4-inch Sears refractor take eyepieces of .95-inch diameter; American scopes, 1.25-inch diameter. What this means is that if and when you move up from a small to a larger scope, you will probably have to start from scratch in the eyepiece department. This is not a big deal: if you sell your old scope, you will certainly sell the eyepieces with it.

Used scopes can be sold—and bought—through the classified columns of *Astronomy* magazine, which include a section on second-hand equipment. I would not, however, *start* with a second-hand scope. Buying a used telescope is by no means as risky as buying a used car, but you still need to know a fair amount about scopes to understand what you are getting for your money.

Some Further Readings

PERIODICALS

Astronomy (monthly), 411 East Mason Street, Milwaukee, Wis. 53202, $15.00 per year, is a fairly new publication aimed primarily at the relative newcomer to stargazing. Its style is generally a good balance between readability and scientific authenticity; other pluses are spectacular color photographs (most of them taken by amateurs) and some excellent articles on sky photography. Its proofreading is sometimes sloppy, and many of its colored paintings and star maps are a lot gaudier than the reality; when they show a "red" star, it's *red!* Also included are monthly star maps, planet positions, etc.

Sky and Telescope (monthly), 49–50 Bay State Road, Cambridge, Mass. 02238, $14.00 per year, is now in its fortieth year of publication. Since it is geared to the fairly sophisticated amateur, most beginners will find it rather heavy going. A notable plus, for readers living in Canada, northern Europe, the southern U.S., Mexico or the West Indies, is its monthly sky map, which is so constructed as to show the stars and horizon at any latitude from 20° North (*e.g.,* Mexico City, San Juan) to 50° North (*i.e.,* all of Canada's cities except those in southern Ontario, plus the U.S. border region west of the Great Lakes). It also runs a supplementary map for Southern Hemisphere readers or travelers, from Latitude 10° South (*e.g.,* Lima, Dar-es-Salaam) to 40° South (*e.g.,* Buenos Aires, Cape Town, Melbourne). The monthly maps in *Astronomy,* by contrast, show only the heavens as seen from Latitude 40° North (*e.g.,* Philadelphia, Columbus, Springfield, Ill., Denver, Salt Lake City), though they are serviceable over most of the United States and southern Canada.

Natural History, published monthly by the American Museum of Natural History in New York, has monthly star maps and a page or so of commentary on celestial current events, but is worth subscribing to only if your main interest is natural history—botany, zoology and anthropology.

Sky Lines, published monthly by the Amateur Astronomers Association, and *Scientific American,* which runs fairly heavy astronomical pieces from time to time, are worth checking out at your local library.

The Astronomical Almanac (annual), Superintendent of Documents, Washington, D.C. 20402, $3.00, contains very detailed information on upcoming celestial events, as does the *Observer's Handbook,* $5.00, published annually by the Royal Astronomical Society of Canada, 252 College Street, Toronto, Ontario M5T 1R7. Some of the same information can also be found in the *World Almanac.*

BOOKS

The following list of readings is by no means definitive, nor have I myself been able to look at all of them (some have been recommended by friends). Readers interested in building up an astronomical bookshelf might well start by checking out the astronomy section of their local library; a quick scan through a book will often tell more about its potential value to the individual reader than any bibliography can.

Star Atlases, Charts and Tables

At least one book of this sort is essential for anyone wanting to go beyond the beginning stages of stargazing. With more detailed charts than could be included in this book, plus more comprehensive tables of interesting objects, a good atlas will give you material for years of watching. Unfortunately most of them, along with this essential material, give you a lot of nonessential information (on telescopes, observing, etc.) which runs up the cost.

Norton's Star Atlas and Telescope Handbook (Sky Publishing Co., Cambridge, Mass.) is one of the standard works in the field. It is obtainable from Edmund Scientific. (See "A Note on Equipment.")

Neale E. Howard's *The Telescope Handbook and Star Atlas* (Thomas Y. Crowell, New York) is similar to the above, but with a rather useful system of transparent overlays enabling you to look at a star chart by itself and then at the galaxies, clusters, etc. in relation to the stars. Like Norton, it's not cheap.

Donald H. Menzel's, *A Field Guide to the Stars and Planets* (Houghton Mifflin, Boston—in paper) is cheaper than either of the above. It has good sectionalized charts of both the stars and the moon, combined with photos which give you some idea of what a particular area actually looks like. Its tables are useful, its monthly star maps and text only marginally so. It also suffers from having been crammed into the pocket-sized format of the Field Guide series— essential for bird-watching, but irrelevant for star-watching.

Cheapest of all is Edmund Scientific's *Mag-5 Star Atlas*, though (as the title suggests) it includes no stars of less than fifth magnitude. I have not checked out its other features, but at $1.00 a copy what can you lose?

Telescope Making

Albert G. Ingalls' three-volume *Amateur Telescope Making*, published by *Scientific American,* is the standard work in the field—but *not* for beginners!

Henry E. Paul's *Telescopes for Skygazing* (Sky Publishing) emphasizes the "assemble it yourself" (*i.e.,* from commercially available parts) approach rather than the do-it-from-scratch approach of Ingalls. N.E. Howard's *Standard Handbook for Telescope Making* (Thomas Y. Crowell, New York) is another fairly elementary manual.

Edmund Scientific carries a line of cheap pamphlets on telescope making by Sam Hall which might be worth sampling; they're only 75 cents apiece.

Sky Photography

Since I know little about this field, I can only suggest you look at *Skyshooting,* by R. N. and M. W. Mayall (Dover, New York), and back files of *Astronomy* and *Sky and Telescope.*

History

The Fabric of the Heavens, by Stephen Toulmin and June Goodfield (Harper, New York), is a literate and sophisticated account of astronomy from its beginnings; it emphasizes intellectual history rather more than "hard" scientific history. A more conventional (and pedestrian) approach is found in Giorgio Abetti's *History of Astronomy* (Abelard-Schuman, New York).

General

The little paperback *Stars* (Golden Press, New York) is good for kiddies, while the same publisher's *The Sky Observer's Guide* is good for older kiddies. It covers some of the same ground as this book.

Robert D. Baker's *Astronomy* (Van Nostrand, Princeton) is a standard text, as is G. O. Abell's *Exploration of the Universe* (Holt, Rinehart and Winston, New York). Somewhat more popular is Anthony E. Fanning's *Planets, Stars and Galaxies* (Dover, New York, in paper).

Nigel Calder's *Violent Universe* (Viking, New York) is a lively layman's guide to the "new" astronomy, including such far-out topics as quasars and black holes; a good buy in paperback. Patrick Moore has published some fifteen titles (Norton, New York) on various aspects of astronomy and telescope making, which should be worth checking out in your library. Robert S. Richardson's *Getting Acquainted with Comets* (McGraw-Hill, New York) is a lively (if occasionally cutesy) guide to a rather specialized field that has attracted many amateurs; among other things, he explains how—with a lot of luck—you might discover your "own" comet, as some amateurs have. Lawrence H. Allen's *Atoms, Stars and Nebulae* (Harvard University Press, Cambridge) will take you into astrophysics—but expect heavy going once you get beyond the early chapters.

Finally, anyone interested in speculating about life on other worlds should by all means read Walter Sullivan's *We Are Not Alone* (McGraw-Hill, New York), the more recent *Intelligent Life in the Universe* (Delta, New York) by Shklovskii and Carl Sagan—a Soviet-American collaboration of two unorthodox but brilliant astronomers—or Sagan and Jerome Abel's *The Cosmic Connection* (Dell, New York, in paper).

Appendix A

Magnitudes

☼ Zero — First
✹ Second
✳ Third
• Fourth
· Fifth

Objects

☆ Open Clusters
✪ Globular Clusters
+ Nebulae
◇ Galaxies
✸ Double Stars

The Greek Alphabet

α	alpha	η	eta	ν	nu	τ	tau
β	beta	θ	theta	ξ	xi	υ	upsilon
γ	gamma	ι	iota	o	omicron	φ	phi
δ	delta	κ	kappa	π	pi	χ	chi
ε	epsilon	λ	lambda	ρ	rho	ψ	psi
ζ	zeta	χ	mu	σ	sigma	ω	omega

The Constellations*

And	Andromeda	Hya	Hydra
Aqr	Aquarius	Lac	Lacerta
Aql	Aquila	Leo	Leo
Ari	Aries	LMi	Leo Minor
Aur	Auriga	Lep	Lepus
Boo	Boötes	Lib	Libra

Cam	Cameleopardalis	Lyn	Lynx
Cnc	Cancer	Lyr	Lyra
CVn	Canes Venatici	Mon	Monoceros
CMa	Canis Major	Oph	Ophiuchus
CMi	Canis Minor	Ori	Orion
Cap	Capricornus	Peg	Pegasus
Cas	Cassiopeia	Per	Perseus
Cen	Centaurus	Psc	Pisces
Cep	Cepheus	PsA	Piscis Austrinus
Cet	Cetus	Pup	Puppis
Col	Columba	Sge	Sagitta
Com	Coma Berenices	Sgr	Sagittarius
CrA	Corona Australis	Sco	Scorpius
CrB	Corona Borealis	Sct	Scutum
Crv	Corvus	Ser	Serpens
Crt	Crater	Sex	Sextans
Cyg	Cygnus	Tau	Taurus
Del	Delphinus	Tri	Triangulum
Dra	Draco	UMa	Ursa Major
Equ	Equuleus	UMi	Ursa Minor
Eri	Eridanus	Vel	Vela
Gem	Gemini	Vir	Virgo
Her	Hercules	Vul	Vulpecula

*Southern constellations not visible north of the tropics have been omitted.

CATALOGUE OF "MARKER" STARS

This list includes all the first-magnitude or brighter stars visible in north temperate latitudes, plus the more useful second-magnitude stars in the same regions of the sky. It can be used in conjunction with the Catalogue of Messier Objects that follows, by the method described in Chapter 5. An interesting point about this list is that while all the stars in question are relatively close—the most distant of them, Rigel, is only 500 light-years off, as compared with (say) 20,000 light-years for the Hercules cluster, M-13—there is otherwise no hard-and-fast relationship between brightness and distance. Thus Alpheratz, though nearly twice as far from us as Alphecca, appears slightly brighter, while Deneb, considerably brighter than either, is nearly ten times as far as Alphecca. The obvious inference is that stars must vary greatly in *intrinsic* brightness; that is, some would be seen to emit far more light than others if all were viewed from the same distance. In fact, we know from many kinds of evidence that this is the case.

Name	Designation	RA h	m	Dec °	,	Mag	Color*	Distance (L-Y)
Aldebaran	αTau	4	33	+16	25	0.9	O	64
Algol	βPer	3	05	+40	46	2-3	B-W	100
Alkaid	ηUMa	13	46	+49	34	1.9	B-W	190
Almach	γAnd	2	01	+42	5	2.3	Y-O	400
Alphecca	αCrB	15	33	+26	53	2.3	W	67
Alpheratz	αAnd	0	06	+28	49	2.2	W	120
Altair	αAql	19	48	+8	44	0.9	W	16
Antares	αSco	16	26	−26	19	0.9	O-R	230
Arcturus	αBoo	14	13	+19	27	−0.1	Y-O	38
Bellatrix	γOri	5	22	+6	18	1.7	B	230
Betelgeuse	αOri	5	53	+7	24	0.4-1.3	O-R	300
Capella	αAur	5	13	+45	57	0.1	Y	46
Caph	βCas	0	07	+58	52	2.4	Y-W	45
Castor	αGem	7	31	+32	0	1.6	W	47
Deneb	αCyg	20	40	+45	6	1.3	W	650
Denebola	βLeo	11	47	+14	51	2.2	W	42
Dubhe	αUMa	11	01	+62	1	2.0	Y-O	105
El Nath	βTau	5	23	+28	34	1.8	B-W	130
Eltanin	αDra	17	55	+51	30	2.4	O	150
Enif	εPeg	21	42	+9	39	2.5	Y-O	250
Fomalhaut	αPsA	22	55	−29	53	1.2	W	23
Hamal	αAri	2	4	+23	14	2.2	Y-O	74
Kaus Aust.	εSgr	18	21	−34	25	2.0	B-W	160
Kochab	βUMi	14	51	+74	22	2.2	O	120
Merak	βUMa	10	59	+56	39	2.4	B-W	76
Mirach	βAnd	1	07	+35	21	2.4	O-R	76
Mirfak	αPer	3	21	+49	41	1.9	Y-W	270
Nunki	αSgr	18	52	−26	22	2.1	B-W	160
Polaris	αUMi	1	49	+89	2	2.1	Y-W	470
Pollux	βGem	7	42	+28	9	1.2	Y-O	33
Procyon	αCMi	7	37	+5	21	0.4	Y-W	11
Rasalhague	αOph	17	33	+12	36	2.1	W	67
Regulus	αLeo	10	06	+12	13	1.9	B-W	78
Rigel	βOri	5	12	−8	15	0.1	B-W	500
Sadr	γCyg	20	20	+40	6	2.3	Y-W	470
Shaula	λSco	17	30	−37	4	1.7	B-W	200
Sirius	αCMa	6	43	−16	39	−1.4	W	8.7
Spica	αVir	13	23	−10	54	0.9	B-W	190
Vega	αLyr	18	35	+38	44	0.0	W	27

*The colors given (B = blue, W = white, Y = yellow, O = orange, R = red) are subtle tints rather than intense colors. A few B-W stars have a slightly greenish cast to some eyes, but this is hard to see. Note also that star colors are a somewhat subjective matter, so that one expert's B is another's B-W, with similar confusions arising between Y-O, O and O-R, etc.

THE MESSIER CATALOGUE

When the eighteenth-century French astronomer Charles Messier compiled his historic catalogue of "nebulae" he was only marginally interested in the objects he was listing. Messier was a comet fancier—a very successful one, since he discovered some twenty during his working life—and his now famous list was merely a by-product of that study. For his own information and that of other comet-hunters, he wanted to note down the various objects which through his small (less than 3-inch aperture) telescope appeared nebulous or hazy, like comets, yet were *not* comets—because they maintained fixed positions in the heavens. In other words, they were objects to disregard.

To today's amateur stargazer, they are anything but that, since the list includes most (though not all) of the "easy" clusters, nebulae and galaxies in the northern heavens. (Why Messier failed to list, say, the Double Cluster in Perseus, while including the Pleiades—which had been known as a group of fixed stars since ancient times—is anyone's guess.) But his list of 103 objects—to which a successor added half a dozen more—has stood up very well; only in six cases did his eyes or his judgment betray him into setting down dubious or fictitious "nebulae."

Some books on astronomy attempt to divide the "Messier Objects" by the season in which they are visible, but this is misleading; nearly all of them can be seen during at least part of the summer, provided you are willing to stay up late enough (or get up early enough). And a numerical listing has advantages of its own, since with it you can quickly find the exact position of an object shown on a map.

The magnitudes listed for these objects require a word of explanation. In dealing with stars, magnitude is an accurate indicator not just of relative brightness but also of relative *visibility*. This is not the case with clusters, galaxies and nebulae, because their total light—which is what magnitude measures—is not concentrated in a point but is spread out over a larger or smaller area, the size of which greatly affects the object's visibility with a given aperture. For example, both the Andromeda Nebula and the Hercules cluster are technically classed as fourth magnitude, yet neither is anything like as visible as any fourth-magnitude star.* Nonetheless, all these objects are visible on a clear night with a 3-inch scope (and not a few, with binoculars)—though knowing exactly which of them you are looking at is something else again. If you can find *and identify* all of the 103 attested Messier objects, you have graduated well beyond the "Beginner" category!

---

*In the case of diffuse and planetary nebulae, magnitude is sufficiently misleading that it is generally not given at all.

Messier No.	RA h.	(1970) m.	Dec °	'	Constel- lation	Visual Magnitude	Type	Comment
1	05	33	+22	00	Tau	—	Diffuse Nebula	Crab Neb.
2	21	32	−00	58	Aqr	5	Globular Cluster	
3	13	41	+28	32	CVn	5	Globular Cluster	
4	16	22	−26	27	Sco	5	Globular Cluster	Unusually large
5	15	17	+02	12	Ser (Caput)	4	Globular Cluster	Spectacular
6	17	38	−32	12	Sco	5	Open Cluster	
7	17	52	−34	48	Sco	5	Open Cluster	Wider than 6
8	18	02	−24	20	Sgr	—	Diffuse Nebula	Lagoon Neb.
9	17	17	−18	29	Oph	5	Globular Cluster	Bright center
10	16	56	−04	04	Oph	5	Globular Cluster	
11	18	49	−06	18	Sct	6	Open Cluster	
12	16	46	−01	54	Oph	6	Globular Cluster	
13	16	41	+36	30	Her	4	Globular Cluster	Hercules Cl.
14	17	36	−03	14	Oph	7	Globular Cluster	
15	21	32	+12	02	Peg	5	Globular Cluster	
16	18	17	−13	47	Ser (Cauda)	6	Open Cluster	
17	18	18	−16	11	Sgr	—	Diffuse Nebula	Omega or Horse- shoe Nebula
18	18	18	−17	08	Sgr	8	Open Cluster	
19	17	01	−26	13	Oph	7	Globular Cluster	
20	18	00	−23	02	Sgr	—	Diffuse Nebula	Trifid Neb.
21	18	03	−22	30	Sgr	7	Open Cluster	
22	18	34	−23	57	Sgr	4	Globular Cluster	
23	17	55	−19	01	Sgr	7	Open Cluster	
24	18	17	−18	26	Sgr	4.6	Open Cluster	
25	18	30	−19	16	Sgr	7	Open Cluster	Sprawling
26	18	44	−09	26	Sct	9	Open Cluster	
27	19	58	+22	38	Vul	8	Planetary Nebula	Dumbbell Neb.
28	18	23	−24	53	Sgr	7	Globular Cluster	
29	20	23	+38	25	Cyg	7	Open Cluster	
30	21	39	−23	20	Cap	6	Globular Cluster	
31	00	41	+41	07	And	4	Spiral Galaxy	Great Neb. in And.
32	00	41	+40	43	And	8	Spherical Galaxy	
33	01	32	+30	30	Tri	8	Spiral Galaxy	Large but dim
34	02	40	+42	39	Per	6	Open Cluster	
35	06	07	+24	20	Gem	5	Open Cluster	
36	05	33	+34	08	Aur	6	Open Cluster	
37	05	50	+32	33	Aur	6	Open Cluster	
38	05	27	+35	49	Aur	7	Open Cluster	
39	21	32	+48	18	Cyg	5	Open Cluster	
40					(2 faint stars taken for Neb.)			
41	06	46	−20	44	CMa	5	Open Cluster	
42	05	34	−05	24	Ori	—	Diffuse Nebula	Great Neb. in Ori
43	05	34	−05	17	Ori	—	Diffuse Nebula	Very close to 42
44	08	38	+19	48	Cnc	4	Open Cluster	Praesepe (Beehive Cl.)
45	03	45	+24	02	Tau	2	Open Cluster	Pleiades
46	07	41	−14	45	Pup	8	Open Cluster	
47					(existence doubtful)			
48					(existence doubtful)			
49	12	28	+08	09	Vir	9	Spiral Galaxy	
50	07	02	−08	18	Mon	6	Open Cluster	
51	13	29	+47	21	CVn	8	Spiral Galaxy	Whirlpool Neb.

52	23	23	+61	26	Cas	7	Open Cluster	
53	13	12	+18	20	Com	7	Globular Cluster	
54	18	53	−30	31	Sgr	7	Globular Cluster	
55	19	38	−31	00	Sgr	4	Globular Cluster	
56	19	16	+30	07	Lyr	9	Globular Cluster	
57	18	53	+33	00	Lyr	9	Planetary Nebula	Ring Nebula
58	12	35	+11	58	Vir	9	Spiral Galaxy	Dim. Close to 59
59	12	41	+11	48	Vir	10	Spherical Galaxy	
60	12	42	+11	43	Vir	10	Spherical Galaxy	Has spiral companion
61	12	20	+04	38	Vir	10	Spiral Galaxy	
62	16	59	−30	05	Oph	7	Globular Cluster	
63	13	15	+42	11	CVn	10	Spiral Galaxy	Bright
64	12	55	+21	41	Com	9	Spiral Galaxy	
65	11	17	+13	17	Leo	9	Spiral Galaxy	
66	11	19	+13	10	Leo	10	Spiral Galaxy	Near 65
67	08	49	+11	55	Cnc	6	Open Cluster	
68	17	38	−26	36	Hya	8	Globular Cluster	
69	18	29	−32	22	Sgr	8	Globular Cluster	
70	18	41	−32	20	Sgr	7.5	Globular Cluster	
71	19	52	+18	36	Sgr	9	Open Cluster	
72	20	52	−12	39	Aqr	8.6	Globular Cluster	
73							(existence doubtful)	
74	01	35	+15	38	Psc	10	Spiral Galaxy	
75	20	04	−22	01	Sgr	8.6	Globular Cluster	
76	01	40	+51	25	Per	12	Planetary Nebula	
77	02	41	−00	09	Cet	10.0	Spiral Galaxy	Small and indistinct
78	05	45	+00	03	Ori	—	Diffuse Nebula	
79	05	23	−24	33	Lep	8	Globular Cluster	Small, bright
80	16	15	−22	55	Sco	7	Globular Cluster	
81	09	54	+69	12	UMa	8	Spiral Galaxy	Great Sp. in UMa
82	09	54	+69	50	UMa	9	Irregular Galaxy	
83	13	35	−29	43	Hya	8	Spiral Galaxy	
84	12	24	+13	03	Vir	11	Spherical Galaxy	
85	12	24	+18	21	Com	10	Spiral Galaxy	
86	12	25	+13	06	Vir	11	Spherical Galaxy	Near 84
87	12	29	+12	33	Vir	11	Spherical Galaxy	
88	12	31	+14	35	Com	11	Spiral Galaxy	
89	12	34	+12	43	Vir	10	Spherical Galaxy	
90	12	34	+13	19	Vir	11	Spiral Galaxy	
91							(Probably a comet)	
92	17	17	+43	11	Her	5	Globular Cluster	
93	07	43	−23	48	Pup	6	Open Cluster	
94	12	50	+41	17	CVn	9	Spiral Galaxy	
95	10	42	+11	52	Leo	10	Spiral Galaxy	
96	10	45	+11	59	Leo	10.4	Spiral Galaxy	Very close to 95
97	11	13	+55	12	UMa	12.0	Planetary Nebula	Owl Nebula
98	12	12	+15	04	Com	11	Spiral Galaxy	
99	12	17	+14	35	Com	10	Spiral Galaxy	Dim
100	12	21	+15	59	Com	11	Spiral Galaxy	Very faint
101	14	02	+54	29	UMa	9	Spiral Galaxy	
102							(Probably same as 101)	
103	01	31	+60	33	Cas	7	Open Cluster	
104	12	38	−11	28	Vir	8	Spiral Galaxy	
105	10	46	+12	45	Leo	11	Spiral Galaxy	
106	12	18	+47	28	CVn	10	Spiral Galaxy	
107	16	31	−12	59	Oph	9	Globular Cluster	
108	11	10	+55	51	UMa	11	Spiral Galaxy	Near Merak
109	11	56	+53	32	UMa	11	Spiral Galaxy	

Appendix B

THE NAMES AND LORE OF CONSTELLATIONS, STARS AND PLANETS

Men have doubtless been seeing pictures in the night sky for as long as they have possessed the imagination to see pictures at all—on the evidence of cave drawings and carvings, at least 30,000 years. From the written records of such ancient peoples as the Egyptians and Babylonians, we know that they gave names to at least some star groupings, though we cannot always be sure which groupings were meant or (even less) that the groupings were the same as our modern constellations. Present-day names for constellations and celestial objects are much more recent. Those of the "northern" constellations—the ones visible from the north temperate zone—are mostly taken from Greek or (occasionally) Roman mythology, as they were listed by Greek and Hellenistic philosophers such as Claudius Ptolemy in the centuries just before and after the birth of Christ. At least some, perhaps all, of these names were of folk origin, coined originally by farmers and shepherds lying out at night under the mild, clear skies of the Mediterranean.

Our names for the "southern" constellations—those visible only from the tropics or farther south—come from a still later period, following the great discoveries of the fifteenth and sixteenth centuries, which is to say, when Europeans had traveled far enough south to see them. Though their names are Latin—the scholarly language of that time—few of them have any mythological counterparts, taking their titles instead from their real or fancied shapes (Crux—the Southern Cross; Tucana—the toucan, etc.).

As regards star names, a few are Latin (*e.g.,* Polaris, Castor, Bellatrix), but most are taken from the Arabic—though (for reasons which will appear) few of these would be recognized as such by modern Arabs. As most people know, when the Arabs—originally a horde of impoverished seminomads—erupted from their desert homeland and grabbed North Africa, most of the Middle East and a large chunk of Spain, they took over much of the learning in the centers of civilization (such as Alexandria and Damascus) they had conquered. Building on these foundations, they became for a few brief centuries the most scientifically advanced civilization in the world outside of China. As part of this scientific advance, they translated classical texts such as Ptolemy's great work on astronomy into their own tongue, including many of the originally Greek names for individual stars. Other stars were listed by purely Arabic names; as desert dwellers, the Arabs had always been very aware of the stars. In the words of their writer Al Biruni, "He whose roof is heaven, who has no other cover, over whom the stars continually rise and set . . . makes the beginning of his affairs and his knowledge of time depend on them."

With the revival of learning in Europe during the Middle Ages, the Arabic translations of such Greek writers as Ptolemy were retranslated into Latin, as were some original Arabic works. The translators, however, for some reason failed to translate the Arabic star names (nearly all of which were intelligible Arabic words) but simply took them over—and garbled them. Most modern star names, therefore, are semiliterate European versions of Arabic names, which in some cases were themselves translations from the Greek.

Greek, Latin or garbled Arabic, star names were devised before the telescope was invented, and hence cover only the brighter stars (and not all of those). The advent of telescopic astronomy forced its practitioners to invent some system of nomenclature to identify the hundreds (eventually, thousands) of objects that the new invention made visible. Beginning with the brightest stars in each constellation (many of which had and have Greco-Arabic names) they designated them by the letters of the Greek alphabet, from Alpha (usually, though not invariably, the brightest star in the group) to Omega. When they ran out of Greek letters, they pressed Latin ones into service, from A to Z, and then double letters, from AA to ZZ. Nowadays, however, most stars beyond the "Greek-letter" group in each constellation are designated by numbers (*e.g.*, 21-Tau; 13-UMa, etc.).

Despite the advance of systematic, scientific nomenclature, however, most astronomers, both professional and amateur, still cling to at least some of the old star names—partly, no doubt, because they are easy to remember, partly because they are more compact (Dubhe, for example, is a lot easier to say than Alpha Ursae Majoris), and partly, I suspect, from the sheer romantic ring of such names as Alpheratz, Antares, Arcturus and Altair—to say nothing of Rasalhague and Zubenelgenubi.

The names of the planets are the simplest of all: Without exception, they are Latin and named for Roman gods (and one goddess)—though Uranus was in fact a Greek god whose name has been Latinized.

I have also included in this list the names of a few asterisms such as the Big Dipper whose lore and folklore is worth writing and reading about. Asterisked names (*e.g.*, *CENTAURUS) are those of stars or constellations visible only from the southern states; most of the rest are visible from the latitude of Washington north, though the most southerly of these may be visible for only a relatively small part of the year. A few, indicated by a double asterisk (*e.g.*, **SCORPIUS) are partly or wholly invisible in Canada and adjacent parts of the U.S.

ADHARA (a-DAR-uh)—Second brightest in CANIS MAJOR, this double star (its main component is pale orange) takes its name from *al adhara*, "the virgins," presumably from some Arabic legend now lost.

ALBIREO (al-bih-RAY-o)—One of the brighter stars in CYGNUS, in which it marks the beak; its name is a landmark in fouled-up philology: The Greek word *ornis* (bird)—another name for the constellation—was first barbarized into Eurism, which then, by a wild leap of imagination, was linked to the plant *erysemon*, the iris—in Medieval Latin, *ireo*. The words *ab ireo* were then further garbled into *albireo* and eventually attached to the star rather than the constellation. Whoever was responsible for this bit of scrambled scholarship eventually turned out to be right for the wrong reasons; Albireo—one of the most beautiful binaries in the sky—is indeed as colorful as the iris, though only through a telescope.

ALCOR—see MIZAR.

ALDEBARAN (al-DEB-a-ron)—Brightest in TAURUS, this ruddy star marks one of the Bull's eyes. Its name comes from Arabic *al dabaran*, "the follower"—*i.e.*, of the Pleiades, since its rising follows shortly after that cluster's appearance in the east.

ALGENIB (al-GAY-nib, al-GEN-nib, al-JEN-ib)—Lying at the southeast corner of the Great Square in PEGASUS, it approximately marks, along with ALPHERATZ, CAPH and POLARIS, the Equinoctial Colure, the celestial Prime Meridian from which right ascensions are measured. Its name comes either from Arabic *al janah*, "the wing," or *al janb*, "the side," both presumably referring to the winged horse of which it is a part.

ALGIEBA (al-JEE-ba)—Second brightest star in LEO, lying in the blade of the SICKLE, and marking the beginning of the mane. Its name is probably from the Latin *juba*, "mane," which somehow became garbled into its Arabic-sounding present form.

ALGOL (al-GOAL)—Second brightest star in PERSEUS, this very noticeable variable is named from the Arabic *ra's al ghul*, the Demon's Head. This, in turn, comes from Ptolemy, who saw it as marking the head of the gorgon Medusa whom Perseus decapitated. It is one of the rather rare class of stars known as eclipsing binaries (see Glossary), the only very conspicuous one of this type.

ALIOTH (AL-lee-oth)—Fifth star in the BIG DIPPER (counting from DUBHE), its name has no plausible derivation. One suggestion, *al yat*, supposedly meaning the fat tail of some celestial sheep, is phonetically possible, but other 'ise unlikely, since the Arabs knew the constellation as the Bear, as we do.

ALKAID (al-ka-EED)—Lying at the tip of the BIG DIPPER's handle, its name comes from *al ka'id banat al na'ash*, "the chief of the daughters of the bier"—the bier meaning the four stars marking the bowl of our dipper. The star is sometimes called Benatnasch, from the same Arabic phrase.

ALMACH (AL-mac)—A well-known binary in ANDROMEDA, its name is said to derive from *al anak al 'ard*, a small badgerlike animal of Arabia. How this creature got into the sky is lost in the mists of time.

ALNILAM (al-NEE-lam, al-NY-lam)—The center (and brightest) star in ORION's belt. Its name is from *al nitham*, "the string of pearls," though the belt more resembles a string of diamonds.

ALNITAK (al-NEE-tak, al-NY-tak)—Easternmost star in the belt, its name is from *al nitak*, "the girdle."

ALPHECCA (al-FEK-a)—Brightest star in CORONA BOREALIS, it takes its name from *al fakkah,* "the dish," this being the constellation's Arabic title.

ALPHERATZ (al-FAIR-ats)—The northwest corner of the GREAT SQUARE, it was originally assigned to PEGASUS but has since been transferred to ANDROMEDA. Its name, however, reflects its original association, being taken from *al surrat al faras,* "the horse's navel." Along with POLARIS, CAPH and ALGENIB (which see) it marks the Equinoctial Colure.

ALSHAIN (al-SHA-in) and TARAZED (ta-ra-ZED)—The two so-called Companions of ALTAIR take their names from *al shahin tara zed,* "the star-striking falcon," a Persian name for the constellation AQUILA in which they lie.

ALTAIR (al-TA-ear, al-TARE)—The brightest star in AQUILA, named from the Arabic *al nasr al tair,* "the flying eagle," the name of the constellation (contrasted with *al nasr al waki,* "the swooping eagle"—see VEGA).

ANDROMEDA (an-DROM-e-da)—Named for the legendary princess, daughter of King CEPHEUS and Queen CASSIOPEIA of Ethiopia, who was rescued from a sea monster by the Greek hero PERSEUS. Appropriately, all four constellations lie adjacent to one another in the northern heavens. Notable for the presence of the GREAT NEBULA (see).

ANTARES (an-TAR-eez)—Chief star in SCORPIUS, its name comes directly from the Greek *anti Ares,* "rival of Ares" (*i.e.,* of Mars). In both its brilliance and its ruddy color, it does indeed rival that planet; a faint bluish companion makes it a binary.

AQUARIUS (a-KWAIR-e-us)—The Water Carrier, one of the twelve zodiacal constellations, is sometimes equated with the demigod Deucalion, central figure in the Greek myth of the Deluge. The basic idea of its name, however, is much earlier; in Babylonian astronomy, this region of the heavens, which includes such other watery constellations as CETUS, DELPHINUS, ERIDANUS and PISCES, was known as "the sea," with Aquarius representing its presiding deity. It is also probably not incidental that the part of the year traditionally assigned to this constellation—January-February—was (and is) the height of the rainy season in the Mediterranean.

AQUILA (ACK-wil-a)—To the Romans, this constellation symbolized the eagle of Jupiter, to the Greeks, "the bird of Zeus," or (in some legends) the eagle that carried Ganymede off to heaven, where he became cupbearer to the gods, or even the vulture that fed on Prometheus' liver.

ARCTURUS (ark-TOUR-us)—The principal star in BOÖTES, its name comes from *arktouros,* "bear guard," one of the Greek names for the constellation (which see). Its conspicuous golden light made it an object of interest in very early times—the more so in that its appearance at sunset in spring heralded the hot, dry Mediterranean summer.

*ARGO NAVIS (AR-go NAV-iss)—Originally the largest constellation in the sky, stretching across a large portion of the heavens south of HYDRA, MONOCEROS and CANIS MAJOR, the Ship has been subdivided for convenience into four: PYXIS (PIX-iss), the Mariner's Compass; PUPPIS (PUP-iss, POOP-iss), the Stern or Poop; VELA (VAY-la), the Sail; and CARINA (ca-REEN-a), the Keel. Of the four, only Pyxis is visible in most of the U.S., while Carina, apart from its brilliant star CANOPUS (which see), is invisible outside the tropics. The Ship itself is, of course, named for the craft in which Jason and his companions sailed in quest of the Golden Fleece. Its missing bow section is said to have been lost on the return voyage when it passed between the legendary wandering rocks in the Hellespont.

ARIES (AH-ree-aze)—Presiding over the zodiacal month March-April, the Ram is traditionally considered the chief sign of the zodiac (in ancient times, the beginning of the year was set in spring). It has also been identified as the ram which carried the Greek hero Phryxus across the Bosporus into Asia Minor, along with his sister Helle; she, however, fell off along the way, into the body of water now known as the Hellespont.

AURIGA (or-REE-ga)—The Charioteer has been seen as symbolizing various chariot drivers in Greek and Roman legend. Certainly the chariot was an important aspect of early (Mycenean) Greek civilization; the mobile warfare it made possible probably helped the takeover of the Greek peninsula from its previous (non-Greek) inhabitants.

BEEHIVE—See PRAESIPE.

BELLATRIX (bel-LAT-trix)—Literally, the Female Warrior or Amazon, its name is said to be a free translation from the Arabic *al najid,* "the (male) conqueror," but how he got to be she is unclear, as is the question of what a woman is doing on the shoulder of ORION.

BETELGEUSE (BET-el-jers)—Brilliant orange, it takes its name from the Arabic *ibit al jauzah,* "the armpit of the Jauzah," the last word being an Arabic name for ORION (which see). With a diameter estimated at 800 million miles, it is one of the largest known stars; if placed in the position of the sun, it would extend far beyond the earth, almost to the orbit of Jupiter. Like all such giant stars, however, it has a very low density.

BOÖTES (bo-OH-teez)—Originally, it is thought, the Ox-Driver (from the Greek *bous,* an ox), but more recently, the Hunter who with his hunting dogs (CANES VENATICI) pursues the bear (URSA MAJOR) round and round the Pole.

CANCER (CAN-ser)—According to one account, it represents the crab that bit Hercules (and was stamped underfoot by him) during his fight with the Hydra. The most obscure of the zodiacal constellations, its most conspicuous object is not a star but the open cluster PRAESIPE. In ancient times, CANCER marked the northernmost point reached by the sun (*i.e.,* on June 21, when it was directly over the Tropic of Cancer), but the slow shift in the earth's axis known as precession has shifted the summer solstice to the border between GEMINI and TAURUS.

CANES VENATICI (CAN-aze ve-NAT-a-chee)—An inconspicuous constellation between BOÖTES and URSA MAJOR, representing the Hunting Dogs of BOÖTES (which see): it is notable for the many galaxies it contains.

**CANIS MAJOR (CAN-iss MAY-jer)—The Greater Dog would be conspicuous, with four second-magnitude stars, even without its centerpiece, SIRIUS, most brilliant star in the sky. It has been associated with various dogs in Greek mythology—most appropriately, with the hound of its neighboring constellation, ORION. In earlier times, The Dog seems to have meant SIRIUS alone (which see).

CANIS MINOR (CAN-iss MY-ner)—An inconspicuous constellation near CANIS MAJOR, except for its principal star, zero-magnitude PROCYON.

*CANOPUS (ca-NO-puss)—Second-brightest star in the heavens (magnitude −0.7), it is also the only naked-eye star in its parent constellation CARINA visible outside the tropics. This constellation was originally part of ARGO NAVIS (the Ship Argo), which see. CANOPUS itself is visible in midwinter from about Latitude 30° (Houston, New Orleans, St. Augustine) on south, as a conspicuous object below and to the right of CANIS MAJOR. The mythological Canopus was chief pilot of the fleet of Menelaus (the husband of Helen), which on its return from the Trojan War touched at Egypt where the pilot died and was commemorated by a monument. Much later, a Greek settlement was founded on the supposed site of his death (near Alexandria) and named for him. (The city has long been in ruins; its site is now occupied by the town of Abu Qir, near where the British Admiral Nelson defeated Napoleon's fleet at the Battle of the Nile.) The legend, however, fails to explain what Canopus is doing aboard the *Argo,* whose legendary voyage would have occurred long before he was born; the name may well be a corruption of some lost word of a different meaning.

CAPELLA (ca-PEL-a)—Literally, "the little she-goat," the star has been associated with various goats in Greek mythology. More likely the name goes back to some earlier, lost legend, or may even have arisen by confusion between the Greek *aigis,* "goat," and *aix,* "storm wind"—Capella, along with its parent constellation, AURIGA, being a herald of the stormy autumn season.

CAPH (KAF)—From *kaff,* "hand," an early Arabic name for its parent constellation, CASSIOPEIA, the reference being to that figure's five conspicuous stars, of which CAPH is the westernmost. With POLARIS and ALPHERATZ it marks the Equinoctial Colure.

CAPRICORNUS (cap-ri-COR-nus)—A zodiacal constellation, its name means either "horned goat" or "goat's horn," but it is usually represented as a semi-aquatic monster—perhaps originally Mesopotamian—with a fish's tail grafted to a goat's head and body. Originally the winter solstice, when the sun stood overhead at the Tropic of Capricorn, lay in this sign, but today it lies on the border between SAGITTARIUS and OPHIUCHUS.

CASSIOPEIA (cas-seo-PEE-a)—This familiar, W-shaped constellation is named for the mother of ANDROMEDA (which see).

CASTOR (KAS-tor)—The dimmer of the TWINS, this conspicuous binary is named for the Roman demigod who with his brother, POLLUX, was a patron of the Roman order of knights *(equites).* The two stars dominate the constellation GEMINI (which see).

*CENTAURUS (cen-TORE-us)—Thought to represent either Chiron, the centaur-tutor of Achilles and other Greek heroes, or some other mythical centaur, this constellation contains Rigil Kentaurus, third brightest star in the heavens and our nearest neighbor in space (apart from the planets)—which, however, is not visible in the U.S.

CEPHEUS (SEF-yoos)—Named for the father of ANDROMEDA (which see), it is one of the circumpolar constellations—*i.e.*, it is visible at any time of year in most of the north temperate zone.

CETUS (SAY-tus)—The Whale is said to represent the sea monster from which PERSEUS rescued ANDROMEDA, turning it to stone by showing it the head of Medusa. This would explain why Cetus lies so far from the other constellations of the Andromeda legend, having sunk well into the south.

COLUMBA (co-LUM-ba)—The Dove is sufficiently inconspicuous (it lies just west of CANIS MAJOR) to have avoided any distinctive name until the sixteenth century; it is supposed to represent the dove Noah released from the ark.

COMA BERENICES (CO-ma be-re-NEE-sez)—The Hair of Berenice, sister and wife of Ptolemy Eugertes, a Hellenistic king of Egypt; according to legend the hair was consecrated as an offering to Aphrodite but was stolen from her temple. Though technically a constellation, it is in fact a nearby and very open cluster. Like CANES VENATICI, it is full of galaxies.

COR CAROLI (COR ca-ROLL-ee)—Though only of third magnitude, this well-known binary in CANES VENATICI is fairly conspicuous for lack of any nearby rivals. Given one of the few relatively modern star names, it represents the Heart of Charles (II), named by some English astronomer as a tribute to that royal profligate and patron of the sciences. Considering Charles' reputation, the star should perhaps be considered sacred to adulterers.

**CORONA AUSTRALIS (co-RO-na aus-TRAL-iss)—This obscure constellation, the Southern Crown, lying just south of SAGITTARIUS, is not as conspicuous as its northern "twin."

CORONA BOREALIS (co-RO-na bo-ree-AL-is)—Like the Southern Crown, the Northern Crown is a rough semicircle of stars (hence their names), but considerably brighter ones, especially the brightest, ALPHECCA.

CORVUS (COR-vus)—Another inconspicuous constellation, the Crow has had that name since the Greeks; it probably goes back to an earlier, folk original. Together with the neighboring constellations, CRATER (the Cup) and HYDRA (the Water Snake), it has been woven into a Roman legend which seems to have been contrived especially to "explain" these three constellations. Certainly it is of little interest otherwise—for which reason I do not recount it here.

CRATER (CRAH-tare)—Usually called the Cup, but the literal translation of the Greek *krater* is "mixing bowl"—*i.e.*, for mingling wine and water at a banquet. To our eyes, it looks rather like a goblet, putting it among the minority of constellations that actually resemble their names.

CROSS, NORTHERN—An asterism lying within CYGNUS. Its upright stretches from DENEB to ALBIREO, with the cross-piece approximately at right-angles; SADR marks the junction. (The Southern Cross, far more conspicuous, is not visible much north of the equator.)

CYGNUS (SIG-nus)—The Swan has not always been a swan, but has generally been classed as a bird of some sort; the Arabs called it a hen, which it certainly resembles far less than it does a swan. One of the more conspicuous summer constellations, it has been equated with half a dozen mythical swans, including Leda's (see GEMINI).

DELPHINUS (del-FEEN-us)—The Dolphin, named for its shape, is conspicuous only because of the scarcity of other star groups in its neighborhood. The dolphin turns up frequently in Greek and Roman legend, among other things as the prototype of philanthropy, both from its alleged concern for its young (it is, of course, a mammal, not a fish) and from various anecdotes concerning its rescue of human castaways from the sea. Modern scientific studies of dolphins indicate that ancient dolphin folklore of this sort probably had a lot of truth in it; the animal *does* care for its young, may have rescued some drowning humans, and is unquestionably an intelligent, playful and nonaggressive beast. Claims that the creature can talk, however, should be taken with several pounds of ocean salt. The constellation is also notable for the names of two of its stars, Sualocin and Rotanev, which when spelled backward form the name of the Italian astronomer Nicolaus Venator (Latinized from Niccolo Cacciatore—in English, Nick Hunter).

DENEB (DEN-eb)—The brightest star in CYGNUS, it takes its name from the Arabic *al dhanab*, "the tail"—which it indeed marks, regardless of what bird is seen in Cygnus.

DENEBOLA (den-EB-o-la)—Like the preceding, this star also marks a tail, but that of *Leo;* its name comes from Arabic *al dhanab al asad,* "the tail of the lion."

DIAMOND OF VIRGO—An asterism formed by the SPRING TRIANGLE (which see) and COR CAROLI.

DIPPER, BIG—Though it is an asterism, not an official constellation, it is far better known than URSA MAJOR, of which it forms a part; I would guess that for every person who can point out the constellation there are a hundred who know the Dipper. In England its seven stars are more often called Charles' Wain (wagon), and indeed it looks not unlike an old-time two-wheeled cart with the wheels removed. The Charles in question is Charlemagne, whose fame in history and legend presumably won him ownership of the wain. The English and Irish also call it The Plow, which again it resembles; an Irish revolutionary group early in this century had as its banner *The Plough and the Stars,* which Sean O'Casey took as the title for one of his plays. Its conspicuous position near the Pole, no less than its two POINTERS which lead the eye to POLARIS, have given it a strong association with the north; the learned, Latinate adjective *septentrional* (northern) derives from the Dipper's seven (Latin *septem*) stars.

DIPPER, LITTLE—Both less conspicuous and less dipperlike than its big neighbor, this asterism does not figure nearly so widely in folklore; see URSA MINOR.

DIPPER, MILK—An asterism sometimes depicted in SAGITTARIUS; see star map on page 66. The "milk" comes from its location in the Milky Way.

DRACO (DRAC-o)—The Dragon has been known for several thousand years, as dragon, snake or similar slithery monster; in recent centuries it has been linked to the Serpent with which Eve had her troubles. Its winding form and relatively dim stars make it rather hard to see.

DSCHUBBA (DSHUB-ah)—Northernmost of the "claw" stars in SCORPIUS, its name probably comes from *al jabbah,* "the forehead," which is confusing. The reference may be a little clearer when we learn that originally Scorpius included both the present constellation and LIBRA, the latter representing the claws, which would put Dschubba more or less on the creature's forehead—if indeed scorpions can be said to have foreheads.

DUBHE (DOO-bay)—Brightest of the POINTERS, takes its name from *al dubb,* "the bear," its Arabic name in full was *thahr al dubb al akbar,* "the back of the Great Bear."

EL NATH (el NATH)—From the Arabic *el natih,* "the butting one"—appropriately enough, since it tips the northern horn of TAURUS.

ELTANIN (el-TA-nin)—From the Arabic *al ras el tinnin,* "head of the dragon," where it is located; the brightest star in DRACO.

ENIF (EN-if)—From the Arabic *al anf,* "the nose"—*i.e.,* of PEGASUS, in which it is the only conspicuous star apart from those marking the GREAT SQUARE.

EQUULEUS (ay-kwu-LAY-us)—The Foal or Little Horse, an inconspicuous constellation lying between PEGASUS and DELPHINUS, has sometimes been taken to represent Celeris, brother of PEGASUS (which see), or any of various Greek mythological horses. Its brightest star is fourth magnitude.

*ERIDANUS (eh-ri-DAH-nus)—Apart from brilliant Achernar, a bright first-magnitude star not visible in the U.S., The River has no star brighter than third magnitude; it winds from near RIGEL deep into the southern hemisphere of the heavens. It is named for a mythological river at whose mouth lay the Electrides or Amber Islands, later identified with the Po river in northern Italy. A more likely original was some river of northern Europe (perhaps the Oder) emptying into the Baltic, whence amber was traded south to the Mediterranean even in prehistoric times. The region of the Po delta might well have been the southern terminus of the amber route across central Europe.

FOMALHAUT (FOAM-al-hout)—The only conspicuous star in its parent constellation, PISCIS AUSTRALIS, (or AUSTRINUS), it takes its name from the Arabic *fum al hut,* "the fish's mouth."

GALAXY, THE (GAL-ax-ee)—The Milky Way takes its name from the Greek *galaxios,* "milky," but was also known to the Greeks (as to many other ancient peoples) as The River of Heaven. Northern peoples were more likely to refer to it as a road or street of some god or other; the Anglo-Saxons for some reason called it Watling Street, after the ancient highway—much of it still in use—from Dover through London to Chester. Not until Galileo was its "milk" identified as tens of thousands of stars too distant to be resolved by the eye; subsequently, it has given its name to the many similar masses of stars known in the heavens.

GEMINI (GEM-i-nee)—The zodiacal constellation of The Twins, who are of course CASTOR and POLLUX. The two stars, close to one another and of almost equal brightness (1.2 and 1.7 magnitude) have been identified as a pair by most ancient peoples, though not necessarily a pair of men. The Greek Twins were hatched, with their sister Helen, from the egg laid by Leda after she had consorted with Zeus in the guise of a swan (CYGNUS, by some accounts). They subsequently played a fairly prominent role in the Trojan War which had allegedly been started by their sister's infidelity—more likely, by the Achaian Greeks' appetite for looting their neighbors. The old-fashioned expression "By jiminy!" is thought by some to have originated as "By Gemini!"

GREAT NEBULA—Refers either to M-31, the well-known galaxy in ANDROMEDA, (the only naked-eye galaxy in the northern hemisphere), or to M-43, the diffuse nebula in ORION—the most conspicuous object of this type visible in the United States.

GREAT SQUARE—An asterism that forms a conspicuous sky mark in late summer; it is composed of ALPHERATZ, in Andromeda, plus ALGENIB and two other stars in PERSEUS.

GUARDIANS OF THE POLE—KOCHAB and gamma-UMi, which form the front of the LITTLE DIPPER.

HAMAL (ham-MOLL)—The brightest star in ARIES, but still only 2.3 magnitude. Its name is from the Arabic *al ras al hamal,* "the head of the sheep"—the sheep, of course, being the constellation itself.

HERCULES (HUR-kyew-lees)—The mythological prototype of this constellation is so well known as to need no description. Oddly enough, however, the Greeks themselves did not originally associate the constellation with their famous hero, calling it instead "The Kneeler." Unlike either ORION or BOÖTES, the other two celestial heroes, HERCULES contains no very conspicuous stars, its brightest being of only third magnitude. Its most notable object is the great globular cluster, M-13.

HYADES, THE (HY-a-deez)—In Greek mythology, the daughters of Atlas, the titan who held up the world, and the demi-goddess Aethra, a daughter of Oceanus; thus, half-sisters of the PLEIADES. Their name, however, means "the rainers," since their appearance in the east at nightfall in autumn and setting at nightfall in spring marked the beginning and end of the winter rainy season around the Mediterranean. The Roman author Pliny wrote of them as "violent and tempestuous . . . causing storms and tempests raging on both land and sea," a judgment echoed in Tennyson's "Thro' scudding drifts the rainy Hyades vext the dim sea." To the Greeks, "Hyades" meant only the six or seven brightest members of this open cluster in the head of TAURUS; we now know that it includes several dozen dimmer stars.

*HYDRA (HY-dra)—The Water Snake winds sinuously southward from CANCER, inconspicuous apart from the second-magnitude star Alphard. It seems to have been known to most ancient peoples as a snake or similar creature; according to one account it represents the dragon that guarded the Golden Fleece, stolen by the Argonauts.

JUPITER (JU-pi-ter)—The title of this planet, like that of VENUS, has no rationale that I have been able to discover. Its name—that of the chief of the Roman pantheon—is appropriate enough for the largest of the planets, but this fact could hardly have been known to the ancients. More relevant, perhaps, is the fact that it can be seen as, in a sense, dominating the heavens; it is brighter than any planet but VENUS and, unlike that body, which as a morning or evening star dances attendance on the sun, may be visible at any time of night. Finally, its measured progress through the signs of the zodiac, almost exactly one sign per year, suggests some sort of "supervisory" function. Its progress through the zodiac is the basis for the twelve-year Chinese (and Southeast Asian) calendar cycle, in which years are named for the Rat, the Dog, the Monkey, etc.—these being Chinese names for the zodiacal constellations through which Jupiter moves.

KAUS (COWSS)—The name of three stars that form the bow of SAGITTARIUS, from the Arabic *kaus*, "bow." Brightest is Kaus Australis, the second word meaning "southern"; similarly Kaus Media and Kaus Borealis mark the middle and northern parts of the bow respectively.

KEYSTONE, THE—A fairly conspicuous asterism in HERCULES, of which hero it supposedly marks the trunk. The cluster M-13 is located on the Keystone's western side. The name is, of course, from the figure's shape.

KOCHAB (KO-chab)—Brightest of the Guardians of the Pole, its name comes from the Arabic *al kaukab al shamaliyy*, "the star of the north"—one of their names for POLARIS. Bar Kochba, the name taken by the leader of the last Jewish revolt against Rome, means "son of a star," and some coins minted during his brief rule bore the image of a star, as a pun on his name.

LACERTA (la-SER-ta)—The Lizard is an obscure constellation between CEPHEUS and PEGASUS, named by the seventeenth-century

Polish astronomer Hevelius apparently as a way of labeling a group of stars not important enough to have been previously named.

LEO (LEE-o)—The Lion is one of the more prominent zodiacal constellations, with first-magnitude REGULUS and second-magnitude ALGIEBA and DENEBOLA. Its name probably comes from its shape, which with some imagination appears as a fairly recognizable lion. To the Greeks, it was the Nemean lion slain by Hercules as one of his labors.

LEO MINOR—Like LACERTA, the Lesser Lion was invented by Hevelius, as a way of tidying up a portion of the heavens between LEO and URSA MAJOR; its brightest star is fourth magnitude.

LEPUS (LEP-us, LAY-pus)—Lying just south of ORION, the Hare is said to have been named by Greek colonists in Sicily, who found these animals a nuisance. Subsequently, the legend grew up that its placement in the heavens was for the convenience of its hunter, ORION, though one would expect that mighty hero to go in for larger game than hares. Its brightest star is third magnitude.

LIBRA (LEE-bra)—To the Greeks, this was not a distinct constellation, but rather the claws of SCORPIUS, its neighbor in the zodiac. Its conversion into a separate sign was presumably motivated by the need for twelve distinct zodiacal figures corresponding to the twelve months. It was thereupon named the Yoke (*i.e.*, the beam of a weighing balance) and eventually Libra, the Balance, but for many centuries continued to be known also as the Claws. Since Roman times, at least, it has been associated with the equinoxes, balancing their equal days and nights.

LOZENGE—An asterism of four stars, lying within a radius of about 5°, which form the head of DRACO; the brightest is ELTANIN.

LYNX—An obscure constellation, lying between URSA MAJOR on the north and GEMINI and AURIGA on the south. Its name is another invention of Hevelius (see LACERTA), who remarked that one would need to be lynx-eyed to make it out. He was right.

LYRA (LIE-ra)—A small but easily found constellation, thanks to brilliant VEGA, the Lyre was named by the Greeks, who saw it as the musical instrument invented by Hermes and by him given to Apollo; for the same reason it was occasionally known as the Tortoise, from whose shell the first lyre was said to have been constructed.

MARKAB (MAR-kab)—One of the brighter stars in PEGASUS, it

forms with ALGENIB the southern side of the GREAT SQUARE; its name comes directly from an Arabic word meaning anything ridden upon—which Pegasus certainly was.

MARS (MARZ)—Its association with the Roman war-god was undoubtedly suggested by its conspicuously ruddy color, though as we now know this comes not from blood but from oxides of iron in its surface rocks, sand and dust. Its two tiny moons (invisible outside an observatory) are named Deimos and Phobos, or Terror and Fear—fit companions for the god of battles. As the only planet with much similarity to earth—notably, its possession of what appear to be ice-caps at its poles—it has for nearly a century been seen by imaginative writers as a possible abode of intelligent life, to the point where Orson Welles' famous radio dramatization of a fictitious Martian invasion in 1938 threw the entire U.S. east coast into a tizzy. One source of the myth, certainly, was the claim of the astronomer Percival Lowell that the planet's surface was covered with a network of canals, presumably made by intelligent beings. More recent observations, especially those from space satellites, have made clear that the "canals" were the product of Lowell's imagination, stimulated, perhaps, by an Italian astronomer's reference to *canali*—which meant merely "channels." Oddly enough, however, satellite photos of Mars have indeed revealed "channels" apparently carved by running water, which so far as we can tell does not now exist on its surface. But there is no evidence of any advanced form of life—and thus far, at least, no convincing evidence for the presence of even primitive organisms such as lichens.

MEGREZ (MEG-rez)—Located where the BIG DIPPER's handle joins its bowl, this star takes its name from the Arabic *al maghrez*, "the root of the tail," the handle also being the very un-bearlike tail of the Great Bear.

MERAK (MARE-ak)—The dimmer of the two POINTERS, and the farthest from POLARIS, its name comes from the Arabic *al marakk*, "the loin" (*i.e.*, of the Bear).

MERCURY (MUR-cure-ee)—Its rapid motion in the heavens compared with that of the other planets made for an obvious association with the Roman messenger of the gods; its elusive quality—it is by all odds the most difficult of the anciently known planets to see—reminds us that Mercury was also the celestial patron of thieves. Satellite observations have revealed that the planet's surface resembles that of the moon, though thanks to its closeness to the sun it is much hotter.

MINTAKA (min-TOK-a)—Westernmost star in ORION's belt, its name is from the Arabic *mintakah*, meaning—"belt." It lies within half a degree of the celestial Equator.

MIRA CETI (MERE-a SAY-ti)—This "miraculous" star in CETUS is one of the most remarkable variables in the heavens, changing from tenth magnitude at its minimum to nearly first magnitude at maximum, so that to the naked eye it seems to appear and disappear. Its period from maximum to maximum averages a little less than a year, but is somewhat irregular, as are its variations in brightness.

MIRACH (MERE-ack)—Its name is apparently garbled from the Arabic *mi'zar*, a girdle or waistcloth—*i.e.*, of ANDROMEDA. The garbling is fortunate in that without it the star could be too easily confused with MIZAR in URSA MAJOR, unfortunate in that it did not go far enough, since the star is still easily confused with MERAK or MIRFAK.

MIRFAK (MUR-fak)—Originally known to the Arabs as *Algenib*, like the star of that name in PEGASUS, it is now known—fortunately— by the distinctive name derived from *al marfik*, "the elbow"— *i.e.*, of PERSEUS, in which it is the brightest star. The brilliant star field surrounding it makes it a specially beautiful object in binoculars.

MIZAR (MEE-zar)—Apparently derived from the same word as the original name of MIRACH (which see), the name seems quite inappropriate, since no bear ever possessed a girdle. Lying next to the end of the Dipper's handle, it is notable for a number of reasons, not least the fact that it was the first star identified as a double through the telescope (in 1650). It is not merely a double but a true binary, both of whose components are spectroscopic binaries (see Glossary). Its faint companion ALCOR (probably from *al khawwar*, "the faint one") has long been regarded as a test of vision, since fairly sharp eyes are required to distinguish it from Mizar. Together the two form the Horse and Rider or (to some American Indians) the Squaw with the Papoose on Her Back. Alcor begot the medieval Latin catchphrase *vidit Alcor at non lunam plenam*, "he sees Alcor but not the full moon"—meaning a person sensitive to trivia but not to important facts.

MONOCEROS (mo-no-SARE-us)—The Unicorn, lying between CAN-

IS MAJOR and CANIS MINOR, just east of ORION, seems to have been identified as a constellation only in relatively modern times, though the name may have been known to the Persians some centuries earlier. Traditionally, the unicorn could only be caught by using a virgin as bait—which may explain why today there are so few of them in captivity; at any rate, MONOCEROS is well separated from VIRGO in the heavens.

NEPTUNE (NEP-tune)—The most distant of the planets visible through a small telescope, its name, that of the Roman ocean god, was probably suggested by its distinctly bluish color and its featureless surface (*i.e.*, as contrasted with that of SATURN or JUPITER). Invisible to the naked eye, it was discovered in the nineteenth century through careful measurements of the motions of URANUS (which see), irregularities in which pointed to the gravitational influence of some unknown body. It is known to have at least two moons.

NUNKI (NUN-ki, NUNG-kee)—Traditionally marking the hand of SAGITTARIUS holding the arrow, its name apparently comes from the Babylonian; I have, however, been unable to discover what it means.

OPHIUCHUS (o-fee-OO-cus)—The Serpent Holder, holding in his hands SERPENS (which see), was named by the Greeks, to whom he symbolized Asclepius, legendary ship's surgeon of the *Argo* and afterward the patron of the medical profession. He was said to be so skilled as to be able to restore the dead to life, for which reason Hades (Pluto), lord of the dead, fearing the loss of his subjects, had Zeus strike him with a thunderbolt and place him in the heavens. The caduceus or staff of Asclepius, entwined with two snakes, is still associated with physicians, notably as the emblem of the U.S. Army Medical Corps.

ORION (oh-RY-un)— The giant hunter and demigod is the most conspicuous constellation in the winter skies, marked as it is by RIGEL and BETELGEUSE, both brighter than first magnitude (though the latter, a variable, drops to first magnitude at its minimum), second-magnitude BELLATRIX and SAIPH, plus the three second-magnitude stars of the belt, not to mention the GREAT NEBULA in the "dagger" hanging below the belt. In Greek mythology, Orion was the lover of the goddess Artemis, who later shot him with one of her arrows—depending on the source, either accidentally or on purpose. Another account has him stung to death by a scorpion, which explains why SCORPIUS is careful to keep out of ORION's way—the two constellations being never visible in the sky simul-

taneously. The three belt stars, whose line points southeast to
SIRIUS, have often received names of their own; in Latin America
they are known as the Three Virgins or Three Kings—the latter
being specially appropriate, since they are prominent in the heavens around Three Kings Day (January 6), when many Latins exchange gifts in commemoration of the gifts of the Magi.

PEGASUS (PEG-a-sus)—The Winged Horse rose from the spring
formed by the blood of the gorgon Medusa who was killed by
PERSEUS. He was captured by the hero Bellerophon, who with his
aid performed many great deeds. At last, Bellerophon attempted
to fly to heaven on Pegasus but fell off; the horse, however,
continued his flight to the place where his constellation still
resides.

PERSEUS (PUR-se-us)—One of the most famous of the Greek
heroes, his great deeds including cutting off the head of Medusa
(see ALGOL) and rescuing ANDROMEDA (which see) from a sea
monster. One of the more conspicuous late summer and autumn
constellations, PERSEUS is notable for its extraordinary double
cluster, visible with the naked eye and spectacular in binoculars
or a telescope.

PHECDA (FEK-da)—Marking the lower left corner of the BIG
DIPPER's bowl, its name comes from the Arabic *al fakhdh*, "the
thigh"—*i.e.*, of the Bear.

PISCES (PIS-kees)—One of the most obscure zodiacal constellations, the Fishes have no star brighter than fourth magnitude.
Their name probably referred originally to some Mesopotamian
fish (or fishlike) god, but to some Greeks they represented Aphrodite and her son, Eros, who leaped into the river Euphrates to escape the monster Typhon.

**PISCIS AUSTRALIS or AUSTRINUS (PIS-kiss aus-TRAL-is, aus-
TRINE-us)—Lying south of CAPRICORN and AQUARIUS, the
Southern Fish, inconspicuous apart from the star FOMALHAUT,
supposedly represents the parent of PISCES.

PLEIADES, THE (PLEE-a-dees)—The Seven Sisters (few people can
see more than six without optical aid) represent the virgin daughters of Atlas, half-sisters of the HYADES and companions of the
goddess Artemis. Pursued along with her by Orion, they were rescued by Zeus, who transformed them into doves *(peleiades)* and
placed them among the stars, where they are known today as one
of the most conspicuous and beautiful open clusters. A telescope
reveals more than a hundred of them.

PLUTO—Discovered only in 1930, this small planet, most distant of the sun's family, was named for the Roman god of the under-world, equivalent to the Greek Hades, lord of the dead. The name was deserved, not merely because Pluto was the only major god not already assigned a planet, but also because the planet, with a surface temperature only a few degrees above absolute zero, is indeed a dead world. Thanks to its small size and enormous distance from earth, not even its diameter is known with certainty. Its orbital period is, of course, the longest among the planets; the last time it lay in its present (1981) position was in 1735. Its orbit is extraordinarily eccentric; at its farthest from earth, it is some 4,700 million miles away, while at its closest, 2,670 million miles, it actually lies within the orbit of Neptune.

POINTERS, THE—DUBHE and MERAK, whose line points (almost) to POLARIS.

POLARIS (po-LA-ris)—Its name comes, of course, from the celestial North Pole, which lies less than a degree from its position. The name, however, is relatively recent; thanks to the phenomenon of precession—a slow shift in the earth's axis with respect to the stars—the star has been located near the pole for only a little over a thousand years, while before that time there had been no true Pole Star for over two thousand years. At various earlier periods Thuban in Draco and KOCHAB had performed the office, while 12,000 years from now the pole star will be VEGA. Since the cycle repeats at intervals of about 26,000 years, in a mere 25,000 years Polaris will again be the Pole Star.

POLLUX (POL-ux)—Brightest of the TWINS (which see), it is named for the hero and demigod who under his Greek name *Polydeukes* figures in the Iliad.

PRAESIPE (pry-SEE-pay)—The Manger (open cluster in CANCER) was traditionally flanked by the Two Asses, the stars gamma- and delta-Cancri, but the source of these names is lost, as is that of its other name, the Beehive. Though a naked eye object, it disappears in hazy weather, whence Pliny's dictum that "if Praesipe is not visible in a clear sky, it is a presage of a violent storm." It was Galileo who discovered that the misty, nebulous object "is not one star only, but a mass of more than forty small stars"; with better telescopes we now can count more than sixty.

PROCYON (pro-SY-ahn)—Its name is from the Greek *prokyon*, "before the dog," since it rises shortly before SIRIUS. With a magnitude of 0.4, it is a fit harbinger for its even more brilliant follower.

*PUPPIS—See ARGO NAVIS.

**PYXIS—See ARGO NAVIS.

RASALHAGUE (RAS-al-hog)—Though only second magnitude, it is still the brightest star in OPHIUCHUS; its name is from the Arabic *ras al hawwa,* "head of the snake charmer," which it indeed marks.

REGULUS (REG-u-lus)—Brightest star in LEO, and the base of the SICKLE, it has been known since Babylonian times as the King— *sharru* in Akkadian, *basilikos* in Greek, *rex* in Latin—from the belief that it ruled the affairs of the heavens, presumably through its association with the king of beasts (it supposedly marks the lion's heart). Copernicus coined its present name, which means literally "the little king."

RIGEL (REE-jel)—Brightest star in showy ORION, it takes its name from the Arabic *rijl al jauzah,* "leg of the giant (or central one)" (see BETELGEUSE).

SADR (SODR)—Lying at the crossing point of the NORTHERN CROSS, its name is from the Arabic *al sadr,* "the breast," where it is located, whether one sees its parent constellation, CYGNUS, as a swan or, like the Arabs, a hen.

SAGITTA (sa-JIT-a)—The Arrow, both small and inconspicuous, lies near AQUILA, with CYGNUS lying opposite. It has been identified with various mythological arrows, including one of Cupid's; by one account, it was shot by HERCULES—not very accurately, it seems, since it misses both Aquila and Cygnus.

**SAGITTARIUS (saj-i-TARE-i-us)—The Archer is usually pictured as a centaur, and in ancient times was often confused with CENTAURUS; it may originally have personified a Mesopotamian war god. Since it lies in the direction of the galactic center, it is notably rich in nebulae and clusters.

SAIPH (SA-eef)—Marking the right knee of ORION, its name was apparently taken from that of some other star in the constellation, since it comes from the Arabic *el saif,* "the sword." The name would much more appropriately be applied to iota-Orionis, lying in the Dagger.

SATURN—Resembling JUPITER to the naked eye but dimmer, it was appropriately named for the father of that mighty god. SATURN was also identified with the Greek Kronos, the god of time, a concept which is perhaps echoed by the planet's leisurely progress through the zodiac—its orbital period is two and a half times that of Jupiter, or nearly thirty years. It was, of course, Galileo, who discovered its spectacular rings, though his telescope was not good enough to give him a clear picture of their structure. Its family of 11 moons is second only to Jupiter's retinue of 14.

SCHEDAR (SHAY-dar)—Brightest star in CASSIOPEIA, its name comes from the Arabic *al sadr,* "the breast," which it marks.

**SCORPIUS (SKOR-pi-us)—The Scorpion either shuns ORION (which see) or is shunned by him. With the addition of LIBRA (which see), it forms a quite plausible picture of a scorpion, with ruddy ANTARES at its heart. From early times it has been considered an ominous and indeed sinister sign in the zodiac, foretelling storms and wars. Like its neighbor, SAGITTARIUS, it is rich in interesting objects.

SCUTUM (SCOOT-um)—Another invention of Hevelius, the Shield was originally Sobieski's Shield, for John Sobieski, King of Poland, who distinguished himself on several battlefields, notably in breaking the Turkish siege of Vienna in 1683 (Hevelius, as a native of Danzig—modern Gdansk—was at least as much a Pole as he was a German). Lying in the Milky Way between the tail of SERPENS and the head of SAGITTARIUS, it contains no conspicuous stars but a great many inconspicuous ones, as well as many clusters and nebulae.

SERPENS (SER-pens)—Uniquely among the constellations, the Serpent is divided into two separate parts, between which is the Serpent-Holder, OPHIUCHUS; these are known as SERPENS CAPUT and SERPENS CAUDA—respectively, its head and tail. The origins of its name probably go back to Babylon or earlier.

SEXTANS (SEX-tans)—This obscure constellation, between LEO and HYDRA, was whimsically named by Hevelius (see LACERTA) in memory of the sextant with which he made his astronomical observations.

SHAULA (SHAW-la, SHAOW-la)—From the Arabic *al shaulah,* "the sting"—*i.e.,* of SCORPIUS; naturally, it was considered an unlucky star by ancient astrologers.

SICKLE, THE—An asterism in LEO, of which it forms the chest and back of the mane; REGULUS lies at the base of the handle.

SIRIUS (SIH-ri-us)—Conspicuously the most brilliant star in the sky, it takes its name from the Greek *seirios,* "sparkling," but is also known as the Dog Star or simply the Dog. However, it did not, as one might think, take over this name from its parent constellation, CANIS MAJOR; rather, the constellation seems to have been named for the star. The most likely explanation is that the star followed the midsummer sun like a dog—*i.e.,* it became invisible (at dusk) shortly before the Summer Solstice, becoming visible again (at dawn) shortly afterward. (This was the case three or four thousand years ago, but because of precession is no longer true.) To the Romans, the period when the star was invisible or barely visible (July and August) was called the Dog days, and was considered unhealthy—justly so, since many intestinal infections, as well as mosquito-borne diseases such as malaria, do indeed peak during the hot season. To the ancient Egyptians, the dawn reappearance of SIRIUS (which they called something like Sot) marked the beginning of their year. This came about through a coincidence: in the centuries around 2500 B.C., SIRIUS reappeared in the latitude of Egypt early in July—which is to say, at just about the time of the annual Nile flood. This was by all odds the most important event of the Egyptian year, since it not only watered the fields—Egypt was and is virtually rainless—but also fertilized them with the silt it brought down from the Ethiopian and Central African highlands.

Sirius is remarkable in another way: while not usually listed as such, it is a binary, though its companion—only ninth magnitude—is very hard to see because of the great brilliance of Sirius itself. The companion is what is called a "white dwarf" star: though its mass is about that of the sun, its volume is only about one thirty-thousandth as great—meaning that its density is enormous; it has been estimated that one cubic inch of its substance would weigh several tons. Such stars—others of the type are now known—are believed to have been formed when much larger stars "ran out of gas," meaning that having burned up all the nuclear fuel whose energy opposed their immense gravity, their substance collapsed and was compressed into a relatively tiny volume. They continue to shine because their cooling proceeds very slowly, thanks to the very small surface area through which heat can be emitted.

SPICA (SPEAK-a)—Marks the ear (Latin *spicum*) of grain held by VIRGO (which see); it lies only 2° from the Ecliptic.

TAURUS (TORE-us)—The Bull, marked by ALDEBARAN, the PLEIADES and HYADES, has always been one of the most prominent zodiacal constellations. It has been associated with various mythological bulls, including the one in whose guise Zeus courted Europa, the sacred bull for whom the Cretan queen Pasiphaë conceived an unnatural passion and to whom she bore the Minotaur, and even the heifer into whom Io was transformed by Zeus, who gave her name to the Bosporus ("ox-ford") where she crossed into Asia Minor. The bull, however, was worshiped around the eastern Mediterranean from very ancient times; archeologists in Turkey have unearthed shrines with bull skulls dating from 7000 B.C. Significantly, the constellation shows only the animal's head and neck.

TRAPEZIUM, THE—The four tightly grouped stars (they are resolvable only in a telescope) that illuminate the GREAT NEBULA in ORION.

TRAPEZOID, THE—An asterism formed by (from east to west) DENEBOLA, Zosma, ALGIEBA and REGULUS, the four brightest stars in LEO; the latter two also form part of the SICKLE.

TRIANGLE, SPRING—The equilateral triangle formed by ARCTURUS, SPICA and DENEBOLA; a conspicuous asterism and sky mark of spring and early summer.

TRIANGLE, SUMMER—The scalene (all sides unequal) triangle formed by VEGA, DENEB and ALTAIR; a conspicuous asterism and sky mark of summer and early fall.

TRIANGULUM—A small constellation just southeast of ANDROMEDA; its three principal stars (only one is even moderately conspicuous) are arranged in a narrow triangle.

TWINS, THE—Either CASTOR and POLLUX or GEMINI.

URANUS (YOOR-an-us)—The first "new" planet to be discovered, by William Herschel by accident in 1791, and named by him after Uranus (Greek *Ouranos*), father of Saturn (Kronos), as the latter was the father of Jupiter. Like JUPITER, SATURN and NEPTUNE, it is a "giant" planet—*i.e.*, much larger than earth but also less dense; it has five moons, none visible outside an observatory. URANUS itself can, under ideal conditions, be seen with the naked eye—if you know exactly where to look.

URSA MAJOR (ER-sa MAY-jer)—Originally, the Great Bear meant only the BIG DIPPER; though its boundaries have been extended in modern times, its name is a mystery. The Greeks associated it with Callisto, yet another of Zeus' conquests, whom he changed to a she-bear to save her from his wife's quite justified jealousy, later placing her in the heavens when she was slain by Artemis. Bears, however, were, like bulls (see TAURUS), worshiped from early times; indeed European caves have revealed caches of cave-bear skulls that appear to have been shrines of Neanderthal man! It is thus just possible that the Bear was the first of all constellations to receive a name—a conjecture given a slight added plausibility by the fact that the constellation was also so named by some American Indians, who until the arrival of Europeans had had no contact with the rest of the human race for some 25,000 years. URSA MAJOR's prominent position near the pole, making it visible all year north of Latitude 40°, must certainly have given it an equally prominent position in the mythology of almost any people that could see it; the Greek name for "bear" *(arktos)* has given us the word "arctic"—the regions ruled by the Bear.

URSA MINOR—The Lesser Bear, less conspicuous than its Greater neighbor, is equally inconspicuous in mythology. The most plausible derivation of its name is simply from its similarity to the other constellation: If the BIG DIPPER is also the Big Bear, then the LITTLE DIPPER is obviously the Little Bear.

VEGA (VAY-ga, VEE-ga)—Its name, originally Wega, comes from the Arabic *al nasr al waki,* one of the Arabic names for its constellation, LYRA (see ALTAIR).

*VELA—See ARGO NAVIS.

VENUS—Named for the Roman love goddess (the Greek Aphrodite), but for no reason I have been able to discover. Venus, indeed, is visible only near dawn and twilight—the hours when a good deal of lovemaking goes on—but this seems a rather implausible theory; the planet also shows phases like those of the moon (which, because of its monthly cycle, has long been associated with femininity), but this has been known only since Galileo. The most likely theory is that the planet was originally associated with an earlier love goddess such as the Semitic Ishtar. Among most ancient Semites (the Jews excepted) this divinity occupied a far more prominent place than among the rather male chauvinist Greeks, so that the brightest of the planets might well have been taken as her symbol. As most people know, indeed, men have been apparently worshiping some sort of female fertility goddess at least since Cro-Magnon times.

VIRGO (VUR-go)—Not the Virgin Mary, but the Greek *kore* or Maiden —like Aphrodite, a symbol of the female fertility principle (see VENUS). The constellation was seen as personifying either Demeter (the Roman Ceres), a goddess of the harvest, or her daughter, Persephone, who after being abducted by Hades was condemned to spend part of the year beneath the earth and part above it. She thus symbolized the seed grain, which is first buried and then brings forth the harvest; with Demeter she was worshiped at Eleusis and later at Athens, under the name of the Mother and Daughter. Appropriately, VIRGO is prominent in the skies of early summer—the Greek grain-harvest season.

VULPECULA (VUL-pe-KYU-la)—Named by Hevelius (see LACERTA, LYNX and SCUTUM), it occupies the portion of the sky between LYRA and SAGITTA—inconspicuously, since its brightest star is barely fourth magnitude. Though the name means literally "the little fox," Hevelius called it "the Fox and Goose." His reasons, though elaborate, make little sense; the chances are he simply liked the name.

ZUBENELGENUBI (ZU-ben-el-je-NU-bee)—From the Arabic *al zuban al janubiyyah,* "the southern claw"—from the days when LIBRA was merely the claws of SCORPIUS (which see). Its neighboring star, ZU-BENESCHAMALI (ZU-ben-esh-a-MOL-ly), takes its name from a similar Arabic phrase meaning—naturally—"the northern claw."

Glossary

ALTITUDE—The angular distance of an object above the horizon, measured in degrees (°) and minutes (').

APERTURE—The diameter of the main mirror or lens of a telescope or binoculars.

ASTERISM—A distinctive group of stars that is not one of the 88 internationally recognized constellations; it may include parts of several constellations (*e.g.*, the Summer Triangle) but more often does not (*e.g.*, the Big Dipper, which is part of the constellation Ursa Major).

ASTRONOMICAL UNIT—The earth's mean distance from the sun, about 93 million miles. Often used to measure distances within the solar system.

ASTRONOMY—The science of heavenly bodies and the universe. It includes *observational astronomy*, which focuses on the locations, motions and appearance of heavenly bodies, *astrophysics*, which is concerned with the processes that formed and maintain them, and *cosmology*, which considers the origins, evolution and fate of the universe as a whole.

AZIMUTH—Loosely, the "true" bearing of a star (*i.e.*, the compass bearing corrected for magnetic variation); technically, the bearing measured from either due south or due north as a zero point (compass bearings on earth are always measured from due north). Stated in degrees (°) and minutes (').

BINARY STAR—A double star system whose components revolve around one another; more rarely, a similar system containing three or more stars. *Visual binaries* (*e.g.*, Antares, Mizar, Polaris) can be seen as such in a telescope, though not necessarily in a small telescope; *eclipsing binaries* are systems in which one component periodically eclipses the other, totally or partially (*e.g.*, Algol); *spectroscopic binaries* cannot be seen as double even in the largest telescopes, their double nature being inferred from variations in their spectrums (*e.g.*, each of the two visible components of Mizar).

CLUSTER, GLOBULAR—A densely packed sphere of several thousand to several hundred thousand stars, held together by their mutual gravitation. Nearly all globular clusters are relatively close to the center of our galaxy (*i.e.*, as compared with the sun), forming a sort of spherical "halo" around it; they have also been observed in other galaxies. In all cases, their stars seem to be billions of years old—*i.e.*, as old as the galaxy itself.

CLUSTER, OPEN—A looser group of a few dozen to (rarely) a few thousand stars, usually associated with clouds of gas and dust. All known open clusters are, like our sun, located relatively far from the galactic center, in the galaxy's spiral arms; they consist of young stars, with ages ranging from a few million to a few hundred million years.

COLURE, EQUINOCTIAL—The arbitrary zero-point used for measuring RIGHT ASCENSION. It is defined approximately by a line from Polaris through Caph, Alpheratz and Algenib to the celestial South Pole.

COMET—A body of the solar system consisting mainly of water-ammonia-methane ice, rocks and dust; when it approaches the sun, the ice evaporates and is driven outward by the SOLAR WIND, forming the comet's tail.

CONSTELLATION—One of 88 recognized groups of stars, named for some animal, object or mythical figure; also, the defined areas of the heavens within which these groups are located. Thus the cluster M-13 is not technically part of the constellation Hercules, but is nonetheless located in that constellation.

DECLINATION (dec)—The angular distance of an object north (+) or south (−) of the celestial Equator; measured in degrees (°) and minutes ('). Equivalent to latitude on earth.

DEGREE—The unit of angular measurement; 360° make a circle, 90° a right angle.

DOUBLE STAR—Loosely, any two stars whose positions (as seen from earth) are very close; in *visual doubles*, the closeness is only apparent, due to the stars' lying in almost the same direction from earth; a *true double* is a BINARY STAR (which see).

ECLIPTIC—The apparent path of the sun through the heavens. Due to the tilt of the earth's axis and the resulting seasonal shifts in the sun's apparent position, the Ecliptic lies north of the celestial equator during the summer, south of it during the winter, crossing it at the Spring and Fall Equinoxes. Note, however, that this refers to the *daylight* hours; at night, the reverse is true—*i.e.*, in summer north of the equator the Ecliptic at night lies south of the celestial Equator. See also ZODIAC.

ELONGATION—The angular distance of a planet from the sun, as seen from earth. The inferior planets (Venus, Mercury) are most easily seen near maximum elongation, when they are least obscured by the sun's glare or skyglow.

EQUATOR, CELESTIAL—An imaginary circle in the heavens lying directly above the earth's equator, and halfway between the celestial poles. See POLES, CELESTIAL.

EYEPIECE—A lens or (more often) combination of lenses in a telescope or binoculars which magnifies the image formed by the OBJECTIVE (which see); usually several different eyepieces are used with a given telescope to yield different magnifications. Eyepieces are described by their focal length, measured in millimeters, but may also be referred to by their power or degree of magnification, which equals the focal length of the objective divided by that of the eyepiece (*e.g.*, an 18-mm eyepiece used with a 900-mm objective will give a magnification of 900/18, or 50×).

GALAXY—Capitalized, the Milky Way; uncapitalized, any very large collection of stars resembling the Milky Way system. Millions of galaxies have been photographed and charted, but nearly all of them are invisible in a small telescope. Galaxies are classified by shape as *spherical, elliptical, spiral* (*e.g.*, the Milky Way and M-31 in Andromeda), *barred spiral* and *irregular* (*e.g.*, the Magellanic clouds).

GIBBOUS—The moon between first quarter and full, or between full and last quarter, when its face is greater than a semi-circle but less than a full circle.

HORIZON—Technically, an imaginary circle 90° below the ZENITH (which see); in practice, it is seldom visible except at sea, or at high altitudes, being obscured by hills, vegetation or buildings.

LIGHT-YEAR—The distance light travels in a vacuum during a year, nearly six million million miles; used to measure interstellar and intergalactic distances. See also PARSEC.

MAGNITUDE—A system of arbitrary units for expressing the brightness of a star or other celestial object. The system is so arranged that a difference of five magnitudes means a hundred-fold increase (or decrease) in brightness—*i.e.*, a difference of one magnitude corresponds to a change in brightness of 2.512 times. The *lower* the magnitude the *brighter* the object; the brightest (*e.g.*, Sirius, Venus) have negative magnitudes.

METEOR—A piece of rock or metal that has entered the earth's atmosphere from space and has been heated by friction to incandescence.

METEOR SHOWER—A display of meteors related to one another through their apparent origin at a particular point in the heavens; many meteor showers occur regularly at particular times of the year (*e.g.*, the Perseids in the period around August 11–12).

METEORITE—The remains of a meteor that has reached the earth's surface (most meteors do not, being burned up in the atmosphere).

NEBULA—Originally, any body that appeared nebulous or cloudy in a telescope; now, clouds of gas and/or dust, usually illuminated by nearby stars. *Galactic* or *diffuse nebulae* are clouds of irregular form, usually associated with open clusters or smaller groups of stars which illuminate them; *planetary nebulae* are much smaller and of roughly spherical shape, surrounding a single star; they are thought to represent debris ejected by the stellar explosions when the stars were transformed into NOVAS (which see) or supernovas. *Dark nebulae* are unilluminated clouds that can be seen silhouetted against nearby diffuse nebulae, or against the Milky Way.

NOVA—Literally, a "new" star, but actually a star that has suddenly increased in brightness to a very conspicuous degree; if the star was previously invisible to the naked eye, it appears as a "new" object. In ordinary novas, the increase in brightness is several thousand times; they are observed from fifty to one hundred times a year in our own galaxy, as well as in other, nearby galaxies. Some ordinary novas are recurrent, making them in effect variable stars of an extreme type. Much rarer are *supernovas*, in which the increase in brightness may be up to 100,000 times; they occur only about once in 400 years in our galaxy, so that much of what we know about them comes from observations of supernovas in other galaxies. The causes that produce novas of both types are not wholly understood. See also NEBULA.

OBJECTIVE—The light-gathering element of a telescope or binoculars.

OPPOSITION—Located on the opposite side of earth from the sun; said of a planet. The inferior planets (Venus, Mercury) never reach opposition, since they are always on the same side of earth as the sun; the other, superior planets lie closest to earth at opposition, hence are most clearly observed at that time.

PARALLAX—The apparent shift in the position of a star when seen from different points on the earth's orbit; more specifically, one half the shift observed from opposite ends of the orbit. Only a few hundred stars are close enough to earth for accurate measurements of their parallax; they are then used as "yardsticks" to estimate the distances of other, more distant stars, based on the different spectroscopic characteristics of their light.

PARSEC—The distance from earth to a star with a *par*allax of one *sec*ond. Parsecs are used as an alternative unit to light-years in stating interstellar distances, with one parsec equaling about 3¼ light-years.

PLANET—A large, spherical object revolving around a star and shining by reflected light. Only the nine planets of our solar system have actually been observed, but close observations of several nearby stars have revealed irregularities in their motions indicating that they, too, possess planets. At one time, planetary systems were thought to be very rare objects in the universe; today they are considered to be relatively common, though this is still far from certain.

POLE, CELESTIAL—A point in the heavens directly over one of the earth's poles.

REFLECTOR—See TELESCOPE.

REFRACTOR—See TELESCOPE.

REVOLUTION—The orbital motion of a body around its "primary" (a planet around the sun, a moon around its planet) or of one BINARY around another.

RIFT—An apparent split in the Milky Way; actually, an elongated dark cloud of dust that obscures the stars behind it.

RIGHT ASCENSION (RA)—The east-west position of a celestial object, comparable to longitude on earth. It is measured in hours (h) and minutes (m) east of an arbitrary zero-line, the EQUINOCTIAL COLURE (which see).

ROTATION—The motion of a celestial body around its own axis.

SATELLITE—A body that revolves around a planet. Since 1957, the natural satellites (moons) of the solar system have been joined by numbers of artificial satellites, used for communication, observation and other purposes.

SOLAR WIND—Gases driven away from the sun by boiling on its surface; their interaction with the earth's magnetic field produces the aurora. See also COMET.

SPECTRUM—Light dispersed according to color by a prism or other device, usually in a *spectrograph*—an instrument designed for the purpose. The spectra of stars (and a few other celestial objects) give invaluable clues to their temperature and composition. The simplest classification of stellar spectra is based on color, which in turn reflects differences in surface temperature. The types range from O (intense blue, temperature around 50,000°) to S (deep red, temperature around 3,000°. Our own sun is Type G (yellowish, temperature around 6,000°).

STAR—A large ball of glowing gas, heated by nuclear reactions in its interior.

TELESCOPE—A device for gathering and focusing light and enlarging the resulting image, by means of an OBJECTIVE and an EYEPIECE (which see). A *reflecting telescope* or *reflector* has a mirror objective; a *refracting telescope* or *refractor* employs a lens objective, while a *catadioptric telescope* uses an objective including both a lens and a mirror. In all three, the eyepieces consist of a lens or lenses. A pair of binoculars is essentially two small, low-power telescopes set into a single frame.

TERMINATOR—The "shadow line" dividing the illuminated from the unilluminated portion of the lunar (or a planetary) disk.

VARIABLE, VARIABLE STAR—A star whose brightness changes, in the majority of cases by an amount barely perceptible to the naked eye. Most variables brighten and dim in a fixed pattern, which repeats at regular intervals ranging from a few hours to a few years; a few, however, vary irregularly.

ZENITH—The point in the heavens directly overhead.

ZODIAC—Twelve constellations lying along the Ecliptic through which the sun, moon and planets appear to move, as a consequence both of their own motions and of the earth's revolution around the sun.

Index

Names of constellations are in capitals (LYRA), those of asterisms (see Glossary) in italic capitals (*GREAT SQUARE*); stars are listed in small capitals (ARCTURUS) and other objects in italics (*Pleiades,M50*). *Galaxies* (with one exception), *globular clusters* and most *open clusters* are listed by their Messier numbers under these headings; *nebulae* are mostly listed by name alphabetically, but are also cross-referenced under the general heading. All features of the *Moon* are listed under that heading.

Page numbers in **boldface** tell where and how to locate the object in question.